EXPLOSIVE WELDING OF METALS AND ITS APPLICATION

EXPLOSIVE WELDING OF METALS AND ITS APPLICATION

by

BERNARD CROSSLAND

The Queen's University of Belfast

CLARENDON PRESS • OXFORD 1982

Oxford University Press, Walton Street, Oxford OX2 6DP

London Glasgow New York Toronto
Delhi Bombay Calcutta Madras Karachi
Kuala Lumpur Singapore Hong Kong Tokyo
Nairobi Dar es Salaam Cape Town
Melbourne Wellington

and associate companies in
Beirut Berlin Ibadan Mexico City

Published in the United States by
Oxford University Press, New York

British Library Cataloguing in Publication Data

Crossland, Bernard
Explosive welding of metals and its application.—(Oxford series on advanced manufacturing; 2)
1. Explosive welding
I. Title
671.5′2 TS227

ISBN 0-19-859119-5

Library of Congress Cataloging in Publication Data

Crossland, Bernard
Explosive welding of metals and its application.—(Oxford series on advanced manufacturing 2; 1)
Bibliography: p.
Includes index.
1. Explosive welding.
I. Title. II. Series.
TS228.96.C76 671.5′2 81–22324

ISBN 0–19–859119–5 AACR2

Printed in Great Britain by
Thomson Litho Ltd, East Kilbride, Scotland.

CONTENTS

PREFACE

Welding of metals is normally associated in people's minds with fusion welding and in particular the various forms of electric welding and gas welding. Probably the first joining process to be used in man's development was that of soldering or brazing, in which a third lower melting point metal is used to joint two metals with a higher melting point. However, the main welding processes which have been used in the last three or four millenniums, such as cold pressure welding of soft metals or forge welding of wrought iron, are solid phase processes.

During this century other solid phase welding processes have been discovered such as friction welding, ultrasonic welding and diffusion bonding. The most recent process discovered has been that of explosive welding, which though it was noted as a scientific curiosity was not thought of as being a potential commercial process until the late nineteen fifties.

An understanding of explosive welding depends on numerous related disciplines such as explosives, the interaction of explosives with metal surfaces they are in contact with, the behaviour of metals under high velocity oblique impact and the parameters which control the overall process.

The application of explosive welding has unique advantages in joining different metals which would frequently be incompatible, but there are great limitations on the geometries which may be satisfactorily welded due to stress intensification arising from reflection of shock waves from changes in section. Large plate cladding to form composite plates for use in pressure vessel construction and for tube plates in heat exchangers is well established and accepted. Cylindrical configurations such as in tube to tube plate welding and in plugging of tubes in a faulty heat exchanger have been commercially exploited. Other applications such as tube to tube joining, lap welding of sheets and explosive fabrication of wire reinforced composites have also been extensively investigated.

The mechanical and metallurgical properties of explosive welds are important. The strength and quality of the bond between a thin cladded layer and the parent plate is difficult to determine whatever the method of fabrication, and explosive welding is no exception. In general the metallurgical problems are probably not as severe as in fusion welding, but there are problems associated with the passage of shock waves through the flyer and parent plate and the extreme conditions experienced at the interface.

Though explosive welding does not compete with conventional welding processes it is a useful additional process in the welding engineer's armoury. This book is aimed at providing an understanding of this new welding process, and its application in engineering.

The author wishes to thank various companies who have provided information including E.I. Du Pont de Nemours and Co. Inc., Nobel's Explosive Co. Ltd., Yorkshire Imperial Metals, Explosive Fabricators Inc., Nitro Metall AB and the Asahi Chemical Co. Thanks are also due to my collaborators over the years, Dr. A.S. Bahrani and Dr. J.D. Williams, and the ten research students who have worked on various aspects of explosive welding. Lastly this book would never have been completed without the unstinting help of my secretary, Miss E.J. Quinn.

1

INTRODUCTION

Welding can be defined as the process of joining two or more materials, often metallic, by localized coalescence or union across the interface. Though the developments in welding have mainly taken place in the last century and particularly in the last three decades, nevertheless welding in the form of soldering was known at least as long ago as 3000 B.C. (1)[†]. In the relief from Im-Dugud in Mesopotamia (3000 B.C.) the horns of the deer depicted are soft soldered to the relief, while hard soldering with silver solder was used in the fabrication of the silver vase of Entemena (2800 B.C.). Burning together bronze, in which the two parts to be joined are held in a clay mould into which liquid metal is poured, also dates to this general period; the modern equivalent is the thermite welding process used for joining rails or cables *in situ*, in which the liquid metal is produced by the exothermic reaction between the metal oxide and finely divided aluminium.

In the period 1900-1400 B.C. (1) wrought iron was probably first produced by the direct conversion of iron ore by charcoal in an enclosed furnace and this produced what was termed a bloom, which has a spongy texture. The hot bloom was formed into a consolidated mass by forging it under a hammer, which expelled the excess slag and welded together the iron. This is really what is termed forge welding, and it was soon realised that the two pieces of wrought iron could be welded together by heating them and hammering them together. An early example of forge welding is provided by the tiny iron headrest found in Tutankhamen's coffin (1350 B.C.) which appears to be made up of two or more pieces.

It was also discovered very early on that soft metals such as gold could be joined together by hammering at room temperature, or what we would now refer to as cold pressure welding. Many examples of such welds have been found but amongst the earliest are some small gold boxes which can be

---

[†]See Bibliography

seen in the National Museum of Ireland in Dublin, which were fabricated by early iron age Irish craftsmen in the period 1000-500 B.C. Another example is a gold torc from County Roscommon in the Ulster Museum in Belfast, which consists of a twisted square-section gold wire which appears to have been formed from two separate lengths joined by a hammered scarf joint; this dates from a period 1500-1000 B.C.

Forge welding and soldering were the only forms of welding until the nineteenth century when the discovery of electric batteries, generators, acetylene and oxygen led to fusion welding processes. The first process to be discovered was that of arc welding by Auguste de Meritens in 1881 which was further developed by Bernados in 1885. Resistance welding was patented by Eliha Thomson in 1886. Slavianoff and Coffin probably discovered bare wire metal-arc welding independently in the period 1888-1892. In the 1880s oxy-coal gas and oxy-hydrogen blow pipes were developed, but real progress in this direction was only made after the development of the cheap oxygen plant by Linde in 1893 and a torch by Fouché and Picard in 1903, which burnt acetylene and gave a much higher temperature, and this led to an acceptable oxy-acetylene welding process.

During this century tremendous advances have been made and many new processes introduced, such as flux-shielded arc welding, gas-shielded arc welding, percussion welding, resistance-spot welding, electron-beam welding, glow-discharge welding, and laser welding. Most welding processes involve melting of the surfaces of the components to be joined or the melting of a low melting point metal to join two components, but several welding processes have been developed which do not involve melting and these are known as solid phase welding processes. Forge welding is a solid-phase process and in recent decades the age old process of hammering together soft metal has led to cold pressure welding; other recently developed solid state processes are ultrasonic, friction, flash-butt, and explosive welding and diffusion bonding. It is a matter of debate if all these processes are entirely solid state processes; for instance, in friction and flash-butt welding there may possibly be some molten material, but if so it is squeezed out in the forging stage. Houldcroft (2) describes and discusses most of these processes while Tylecote (3) gives greater detail of the solid state processes.

The question which must be answered is: What is welding? An ideal weld is one in which there is complete continuity between the components joined, and every part of the weld is indistinguishable from the metal of the component parts. For the two surfaces to unite it is necessary to bring the two surfaces sufficiently close together to be within the range of the inter-atomic attractive forces. The force between two atoms consists of attractive and repulsive forces as shown in Fig. 1.1, and at a certain equilibrium distance these two

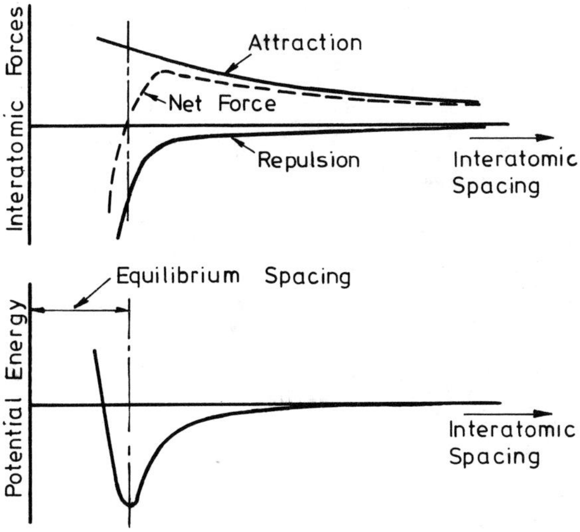

FIG. 1.1 Forces and potential energy between pair of atoms

forces will be in equilibrium or, put another way, the potential energy reaches a minimum value. For two surfaces to adhere they must be brought together to within this equilibrium distance and for this to occur it is necessary for them to be free from oxide and other contaminant films. The strength of the bond formed will depend on the crystallographic misorientation at the interface, and the diffusion and recrystallisation, which are both temperature dependent. The situation in relation to the adhesion between two different metals which may not only have different atomic spacing but also a different crystalline structure, is obviously very complex, but nevertheless, though there may be a distortion of the atomic structure on each side of the interface, there will be an equilibrium distance when the attractive and repulsive interatomic forces are in balance.

The essential problem in joining two metals is how to achieve virgin surfaces without oxide or other contaminant surface film. Such a state of cleanliness is impossible to achieve by normal mechanical or chemical processes, as the surfaces if cleaned are immediately subject to re-oxidation. It is possible to produce the required state of cleanliness under very high vacuum, and if surfaces produced under these conditions are brought together under vacuum then adhesion will occur between the asperities. So essentially welding must provide a way of obtaining intimate contact between the two components to be joined. Basically this can be achieved in two ways, which form the two distinguishable groups of welding processes:

(a) *Fusion Welding*. In fusion welding the two metal surfaces being joined are melted by the application of heat, and the contaminant surface films are either brought to the surface of the melt pool or are dissolved within the molten metal. Many fusion welding processes have been developed which use different sources of heat such as oxy-acetylene, electric-arc, electron-beam and laser. The disadvantages of fusion welding are that the zone joining the two components is basically a cast structure, which suffers defects such as cracking and formation of cooling cavities, and also inferior mechanical properties such as ductility and impact strength. The properties of this cast zone can be improved if a filler rod is used during welding, as this enables beneficial elements to be added to the weld zone. The heat in the melted zone is dissipated partly through the adjoining solid metal and the resulting temperature rise gives a heat-affected zone in which there may be harmful metallurgical changes.

Flow welding may be considered as another form of fusion welding in which a third lower-melting-point metal is used to join two components. The contaminant surface films are either dissolved by the molten metal or they are brought to the surface of the melt pool, though some chemical flux may be required to lower the melting point of the contaminant film or increase its solubility in the molten metal. Soldering and brazing are examples of flow welding.

(b) *Solid-Phase Welding.* Solid-phase welding, as its name suggests, involves no melting of the components. Many solid-phase welding processes involve the application of a pressure sufficient to cause gross plastic deformation at the interface of the components being welded. The plastic deformation increases the area of contact and breaks up the more brittle contaminant surface film, which allows virgin metal to be extruded into contact with the adjoining metal through the gaps created by the breaks in the oxide film, and to form a weld. The process can be carried out cold, as, for example, in the cold pressure welding of soft metals such as copper and aluminium. However, with harder metals the temperature is raised, as in forge welding, as at an elevated temperature metals become more plastic at lower stress levels so that they are more readily deformed. In some cases there is the additional advantage that the hardness of the oxide film relative to that of the base metal increases with temperature, so the oxide film breaks more readily under plastic deformation.

In diffusion bonding the components are pressed together under a modest pressure and at an elevated temperature while under vacuum. The pressure and temperature are such as to cause little, if any, plastic flow and the bond is produced by diffusion across the interface. Sometimes a thin layer of a third metal is introduced between the components.

In forge welding both components are heated prior to forging, but a similar situation can be created locally at the interface between the components by frictional heating, as in friction or ultrasonic welding, or by electric heating, as in resistance-butt welding. In friction welding heat is produced at the interface between the two components by relative rotation under a pressure applied normal to the interface, which produces intense local heating so that the metal close to the interface becomes very plastic. An increase in the end load is then sufficient to cause intense deformation at the interface, and this deformation allows a weld to be created. In ultrasonic welding the heat is produced at the interface between the two components, while under a pressure normal to the surface, by ultrasonic vibration of one surface relative to the other induced by a force transducer. In resistance-butt welding the localised heating at the interface is produced by the contact resistance when an electric current is passed across it.

In explosive welding there is a high-velocity oblique impact between the two components being welded which causes the metals to behave like fluids. As a result a high-velocity jet is formed from the two surfaces of both components, which leaves two virgin clean surfaces which are pressed together to form a weld.

In fusion welding it is difficult to weld metals with greatly differing melting points, and obviously it is impossible if the boiling-point temperature of one is close to the melting-point temperature of the other. Even with some metals with a similar melting point there are serious metallurgical problems, such as the formation of intermetallic compounds with undesirable properties, as in the case of titanium welded to steel, when a very brittle compound is formed. These problems are in addition to the problems associated with a cast structure in the weld zone, and other metallurgical problems which can arise in the heat-affected zone.

Though the metallurgical problems in the solid-phase welding processes are less acute than in fusion welding processes there are other serious limitations. In cold pressure welding very large deformations and pressures are required to obtain adequately strong welds, so the process is only applicable to simple shapes and relatively small areas, and even then only if the large plastic deformations are acceptable. It is used, for example, in joining soft metal cylindrical wires together or obtaining a low-resistance connection between soft metal strips. With friction welding the process is limited to circular sections, with the possible exception of orbital friction welding, and additionally the end force and driving power are a serious limitation on the size of components which can be welded, unless the considerable capital cost involved in such a machine can be economically justified. In diffusion bonding the problem is how to apply the considerable forces required to hold two large components in intimate contact, while they are held at an elevated temperature under a high vacuum. In some cases the interfacial pressure can be generated by the atmospheric pressure acting on the outside of the vacuum system. Finally, with explosive welding only relatively simple shapes can be welded such as plate to plate, as in explosive cladding, or cylindrical surfaces, as in the welding of a tube-end into a tubeplate or the cladding of the inner surface of a tube.

Historically explosive welding was noticed, if not recognised as a solid-state welding process, during the First World War, when high-velocity pieces of shrapnel from the disintegration of the metal casings of shells or bombs were observed to stick to steel stanchions or other metal surfaces. However, there appears to have been no scientific investigation of this phenomenon. It appears that Carl (4) was the first to have noted welding under high-velocity impact in experiments in which a detonator was separated by two half-hard brass shims from a booster charge of high detonation velocity explosive. Recovery of the brass discs after detonation of the charge showed that they had been welded together, and a micrograph of the section shows small poorly formed interfacial waves now frequently associated with explosive welds. Carl concluded that the weld was not a fusion weld but that it had been formed by a solid-state mechanism. Lavrentiev, as reported by Deribas (5), observed explosive welding in 1946, but his concern was with the waves generated at the interface.

In the late nineteen fifties it would appear that several research workers started to study the possibilities of explosive welding independently of one another, and it is difficult to establish who was first to realise the practical value of explosive welding. Credit is frequently given to Philipchuk (6) who in 1957 noted explosive welding when explosively forming an aluminium U-channel in a steel die, but perhaps more importantly he recognised its potential commercial value by filing a patent application in 1960 which was granted in 1962 (7). Pearson (8) (9) (10) (11) was in the period 1956-1958 working on a family of explosive presses for powder compaction, and he noted if the press plates were misaligned so their faces were not parallel, then the plates became explosively welded together. He also observed waves at the interface similar to the surface waves noted by Allen, Mapes and Wilson (12) in their work on projectiles impacting metal targets at oblique incidence. According to a private communication Pearson and Allen concluded that the two phenomena were related to the same basic process, and they used the term 'surface jetting' to describe what was happening in both areas of study. This was perhaps the first appreciation of what is now recognised as probably the basic mechanism of explosive welding. Other early workers in the field were Davenport and Duvall (13), Cowan, Douglass and Holtzman, who filed a patent application in 1960 (14) which demonstrates their understanding of the jetting phenomena, and Tardif (15). It appears that the realisation of the potential commercial value of explosive welding stimulated early work in Russia, Japan and the United Kingdom.

Since these early days explosive welding has become recognised as a practical process and manufacturing facilities exist in the U.S.A., United Kingdom, West Germany, Sweden, Japan, Brazil, Russia, East German Democratic Republic, Poland, Czechoslovakia, Bulgaria and China, and no doubt in other countries. Explosive welding has mainly found commercial application in large plate cladding of one metal on another, tube to tubeplate welding, cladding one tube on another, plugging of heat exchangers, various electrical connectors, especially those between copper and aluminium, transition pieces, especially for pipework in cryogenic systems, pipe to pipe welding, etc. Other applications have been investigated but not exploited, such as welding of a collar onto a tube, high tensile wire reinforced aluminium and other combinations, tubes with axial leak detection channels incorporated in the wall of the tube, combined forming and welding, etc.) During the last fifteen years many papers have been written including review papers by Crossland and his collaborators (16) (17) (18), and several conferences have been held which have produced proceedings, such as those organised by the Denver Research Institute (19), the High Pressure Technology Association (20), the Welding Institute (21), and the Czechoslovakian Institute of Industrial Chemistry (22). Reference will be made to individual papers published in these proceedings and elsewhere in the succeeding chapters.

2

THE MECHANISM OF EXPLOSIVE WELDING

It is generally accepted that explosive welding occurs under high velocity oblique impact, though it is conceivable to use explosive energy to form a conventional cold pressure weld. The basic set-up to achieve welding between two metallic plates is illustrated in Fig. 2.1, in which the top, or 'flyer plate' is supported with the minimum of constraint at a small angle of incidence, α, relative to the stationary, or 'parent plate', which is supported on an anvil which might be a metal plate or more usually sand. The top surface of the flyer plate is covered with a protective layer or 'buffer', which may be rubber or polythene or cardboard or even a thick coat of plastic paint depending on the explosive used. A layer of high explosive, either in the form of a sheet explosive or powder, is laid above the buffer and it is detonated from the lower edge.

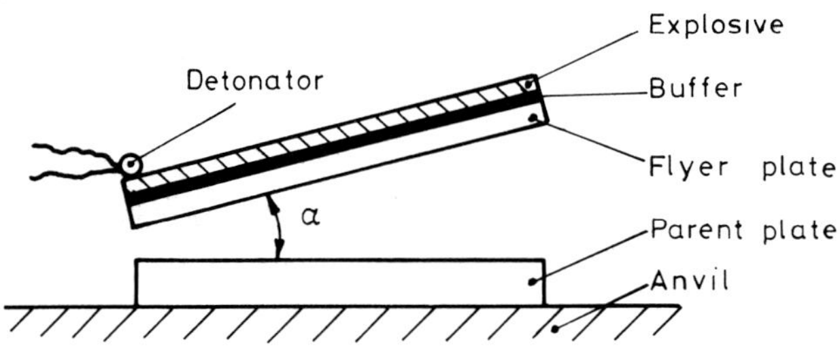

FIG. 2.1 Basic set-up for explosive welding

It is fairly obvious that there are several problems involved in analysing the dynamics of the process. Firstly, it is necessary to understand the interaction between the explosive and the flyer plate. This requires a knowledge of the equation of state for the explosive, but for the explosives commonly used in explosive welding this is not yet established. Secondly, it is necessary to understand the response of the flyer plate to the explosive loading, so as to determine the velocity and deformation of the flyer plate prior to collision with the parent plate. Lastly, it is necessary to understand what occurs during the collision between the flyer and parent plates. It is in this final stage that the major difficulty in understanding occurs, as very high instantaneous pressures are generated with associated very high strain rates, and under these conditions it is near to impossible to define material properties. It is under these circumstances that it is necessary to make some simplifying assumptions in order to begin to understand what is happening.

Birkhoff et al (23) were faced with a similar difficulty in understanding the mechanism of the hollow charge, which it will be seen has some similarities to the mechanism of explosive welding. It was discovered as long ago as the end of the eighteenth century, that the force of an explosive charge could be concentrated on a small area by hollowing out the charge opposite this area, hence the term hollow charge. The effect has become known as the Munroe or Neumann effect. However, it was not until the Second World War that it was discovered that armour plate or concrete walls could be penetrated by a relatively small hollow charge, if the cavity was lined with metal. This formed the basis of the bazooka anti-tank weapon, which could be carried and operated by individual infantry men, and which fired a rocket propelled missile with a hollow charge warhead. For a warhead the cavity in the explosive is conical and it is lined with a metal cone normally of steel or copper, whereas for a line cutting charge the cavity is wedge-shaped as shown in Fig. 2.2a. In the case of the wedge-shaped charge the detonation of the booster charge starts an explosive wave in the main charge, which when it reaches the apex of the wedge causes the metal liner to collapse, as shown in Fig. 2.2b, forming a high velocity metal jet. A similar situation applies to the conical cavity in a warhead, and it is the resulting high velocity jet which has a high penetrative capability when fired against conventional armour plate, and which when it has penetrated the armour plate can cause considerable havoc within the tank. There are obvious similarities between this situation and that in explosive welding, so it is relevant to consider the analysis developed by Birkhoff et al.

The mechanism of the liner collapse for a wedge-shaped cavity is shown in Fig. 2.2. The very high pressure exerted by the detonation wave on the liner causes the liner to collapse

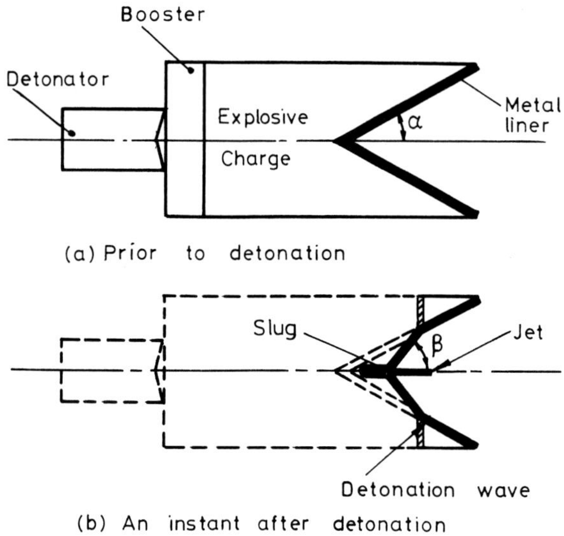

FIG. 2.2 Hollow charge with wedge-shaped liner (23)

The forces are so great that the strength of the liner material can be neglected, and the material may be assumed to behave as an inviscid fluid, though this does not imply that it is necessarily in a fluid state, and indeed there is evidence that it is not. The explosive pressure on the outside of the liner causes it to move inwards at a velocity V_0, in a direction which Birkhoff et al shows bisects the angle between the perpendicular to the original liner surface, and the perpendicular to the collapsing liner surface as shown in Fig. 2.3. The

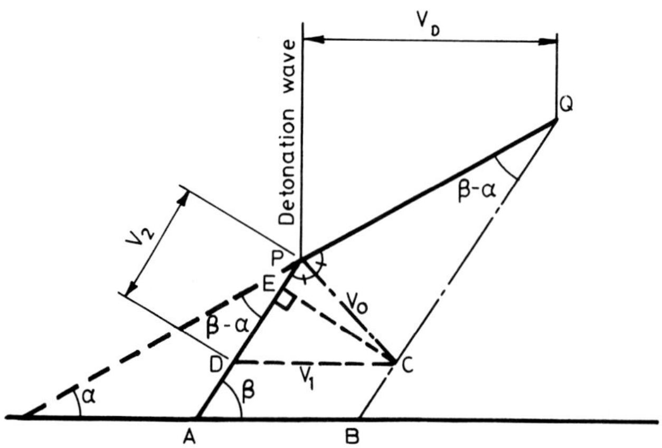

FIG. 2.3 Geometry of the liner collapse process (23)

moving liner retains a wedge shape with the apex moving to the right. To the left behind the moving apex is a section of completely collapsed wedge which, in fact, contains metal from the outer surface of the wedge liner. The inner part of the wedge forms a jet which is squeezed out from the inner apex of the liner and travels at a very high speed along the axis, to the right in Fig. 2.2. In other words the metal in the wedge lining divides into two parts, the outer surface being formed into a relatively slow moving slug, which travels to the right, and the inner surface forming a fast moving jet which travels to the right.

At zero time it will be assumed that the detonation wave has reached the plane passing through point P in Fig. 2.3, and in unit time it will have moved to the plane through Q, so the horizontal distance between these planes will be equal to the detonation velocity V_D. To an observer moving with point P, the undeformed liner material will appear to be moving towards him along QP with a velocity represented by the vector QP, and it will leave him along the line PA. The change in direction of the liner material relative to the observer is caused by the momentum imparted by the pressure in the detonation wave. As the pressure exerted by the detonation wave is everywhere perpendicular to the deforming liner, the velocity of the liner relative to the observer will be the same as it approaches him as when it leaves him. Consequently, the velocity of the liner leaving the observer will be represented by QC = QP, and the closing vector PC is the velocity V_o imparted to the liner by the detonation wave. It will be noted that the triangle CQP is an isosceles triangle, and since QC is parallel to PA the angle PCQ = angle CPQ = angle CPA, and therefore V_o bisects the angle APQ.

At zero time the liner is in the position APQ and after unit time it is in position BQ, so the apex of the wedge has moved from A to B. Considering triangle CPD it can be seen that A moves to B with a velocity:

$$V_1 = \frac{V_o \cos \frac{1}{2}(\beta - \alpha)}{\sin \beta} \qquad (2.1)$$

An observer moving with A would see P coming towards him with a velocity represented by vector DP which will be:

$$V_2 = V_o \left[\frac{\cos \frac{1}{2}(\beta - \alpha)}{\tan \beta} + \sin \tfrac{1}{2}(\beta - \alpha)\right] \tag{2.2}$$

As viewed by the observer at A the process appears to be in a steady state and as the liner material may be assumed as incompressible, then Bernoulli's equation can be applied:

$$p = \tfrac{1}{2}\rho u^2 = \text{const} \tag{2.3}$$

So the pressure at any point in the fluid determines the velocity of the fluid at that point. If it is assumed that the pressure of the exploded gases rapidly decays, then the pressure on the collapsing surfaces remote from P will be very low and can be considered constant. Hence, as viewed by the observer at A the jet and the slug will appear to recede with exactly the same velocity, V_2, as the liner appears to approach, as indicated in Fig. 2.4. It will be observed that the lengths of the jet and the slug will be the same.

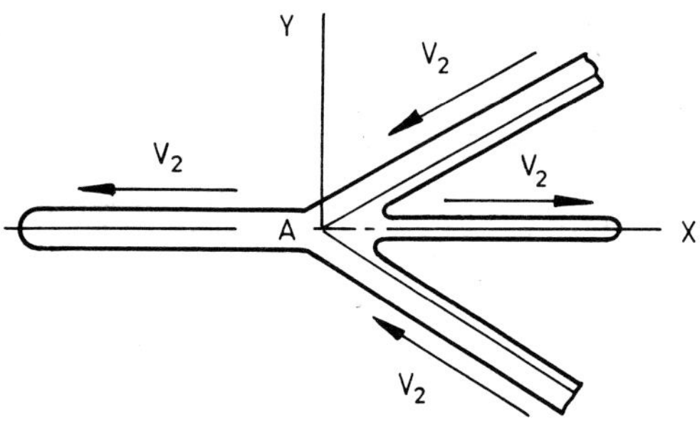

FIG. 2.4 Formation of jet and slug as seen by an observer at the moving junction A. (23)

The absolute velocity of the jet will be:

$$V_j = V_1 + V_2 = V_o \left[\frac{\cos \frac{1}{2}(\beta - \alpha)}{\sin \beta} + \frac{\cos \frac{1}{2}(\beta - \alpha)}{\tan \beta} + \sin \tfrac{1}{2}(\beta - \alpha)\right] \tag{2.4}$$

while the absolute velocity of the slug will be:

$$V_s = V_1 - V_2 = V_o \left[\frac{\cos \frac{1}{2}(\beta - \alpha)}{\sin \beta} - \frac{\cos \frac{1}{2}(\beta - \alpha)}{\tan \beta} - \sin \tfrac{1}{2}(\beta - \alpha)\right] \tag{2.5}$$

During the collision of the walls of the liner momentum is conserved, and this determines the division of mass between the slug and the jet. Referring to Fig. 2.4 it can be readily observed that momentum in the Y direction is conserved, and for it to be conserved in the X direction it is necessary to satisfy the equation:

$$mV_2 \cos \beta = m_s V_2 - m_j V_2 \tag{2.6}$$

where m is the total mass of the liner, m_s is the mass of the slug at the end of the process and m_j the mass which goes into the jet. As there is conservation of mass:

$$m = m_s + m_j \tag{2.7}$$

From 2.6 and 2.7 it can be deduced that:

$$\begin{aligned} m_j &= \frac{m}{2}(1 - \cos \beta) \\ m_s &= \frac{m}{2}(1 + \cos \beta) \end{aligned} \tag{2.8}$$

The magnitude of the velocity imparted to the liner depends on the explosive used and the explosive loading, which is the mass of explosive per unit mass of liner. If V_o is known and the velocity of detonation, then by pure geometry from Fig. 2.3:

$$\frac{V_D}{\cos \alpha} = \frac{V_o \cos \frac{1}{2}(\beta - \alpha)}{\sin (\beta - \alpha)} = \frac{V_o}{2 \sin \frac{1}{2}(\beta - \alpha)} \tag{2.9}$$

For a given α it is possible from 2.9 to deduce β, and knowing β the division of mass between the slug and the jet can be determined. For instance, if $V_D = 7.0$ km/s, $V_o = 1$ km/s and $\alpha = 15^o$, then $\beta = 22.9^o$ and 3.9% of the liner goes into the jet and 96.1% forms the slug. The jet velocity is $V_j = 4.992$ km/s and the slug velocity is 0.133 km/s.

The slug is an example of explosive welding as it is formed from the two sides of the wedge liner. Basically the inner surfaces of the liner have been removed to form the jet, leaving two virgin clean surfaces which are forced together to form a solid state weld. However, in the hollow charge situation the explosive loading is much greater than is necessary for welding, as the objective is to produce an effective jet, and as a result the slug will be very severely damaged. Pearson and Hayes (11) showed the equivalent set-up which they had used for producing explosive welds and this is shown in Fig. 2.5.

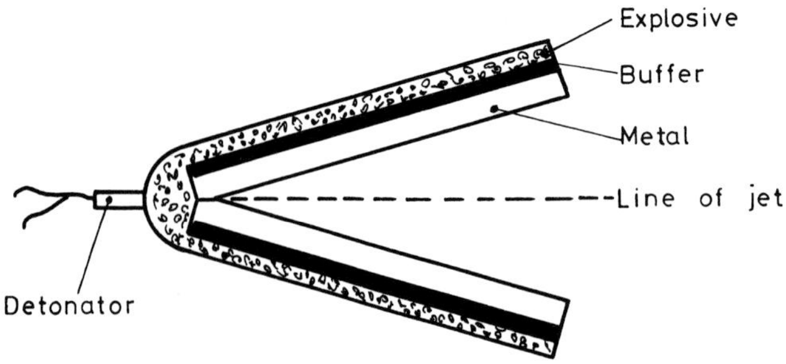

FIG. 2.5 Symmetrical oblique impact welding (11)

The explosive welding set-up shown in Fig. 2.5 has been used, but practically it is not viable for large plate cladding as it poses problems in setting up. By far the most usual arrangement is basically that depicted in Fig. 2.1, though as will become apparent the flyer plate is mostly arranged parallel to the parent plate with a pre-determined stand-off, but this does not affect the analysis. Fig. 2.6a shows the geometry some short instant of time after the detonation has been

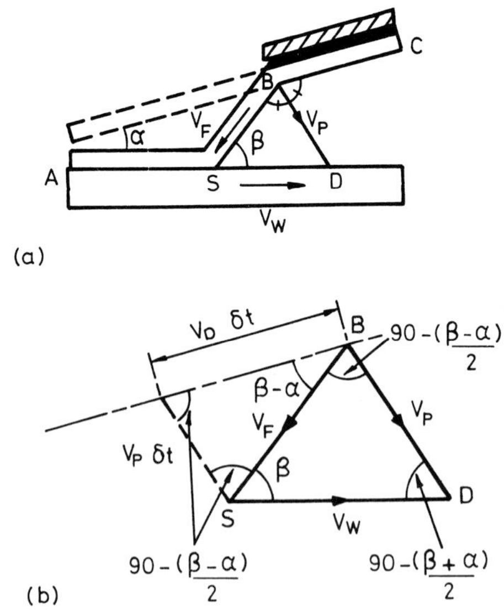

FIG. 2.6 Geometry of collapse of flyer plate

initiated, and before the detonation wave has reached the extremity of the charge. The direction of the velocity of the flyer plate after the detonation wave has passed is not readily deducible, and consequently various assumptions have been made which apparently make little difference to the final results of the analysis. In the present analysis the same argument will be used as that given by Birkhoff et al (23), and consequently the direction of the velocity of the flyer plate bisects the angle SBC. The magnitude of this velocity will depend on the explosive used and the explosive loading.

From the geometry of the vector diagram shown in Fig. 2.6b it can be shown that:

$$\sin \tfrac{1}{2}(\beta - \alpha) = \frac{V_P}{2V_D} \tag{2.10}$$

from which knowing V_P, V_D and α it is possible to deduce the dynamic angle of obliquity, β. Considering the triangle SBD then:

$$V_W = V_P \frac{\cos \tfrac{1}{2}(\beta - \alpha)}{\sin \beta} \tag{2.11}$$

and

$$V_F = V_P \frac{\cos \tfrac{1}{2}(\beta + \alpha)}{\sin \beta} \tag{2.12}$$

where V_W is the velocity of weld propagation and V_F is the velocity of the flyer plate relative to an observer moving with S. It will be noted from Equations 2.11 and 2.12 that there is very little difference between V_W and V_F for small values of α, and no difference for the parallel set-up when α is zero.

To an observer at S the process appears to be in a steady state, and to preserve momentum in the horizontal plane a backward or 're-entrant jet' must be formed, while part of the flyer plate forms a 'salient jet' as shown in Fig. 2.7.

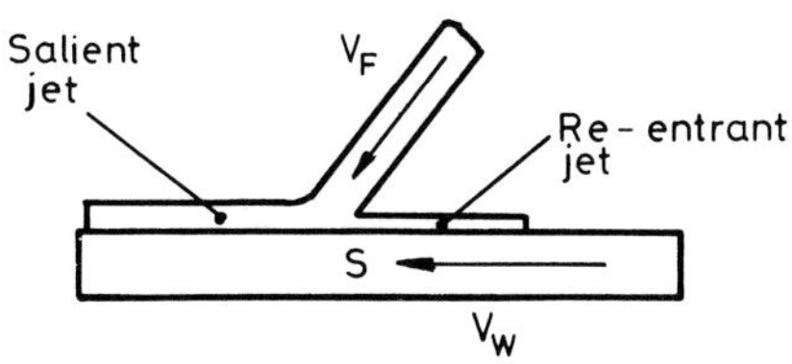

FIG. 2.7 Formation of re-entrant jet

From conservation of mass and momentum it can be shown, as for the hollow charge, that the division of mass between the salient and re-entrant jets is:

$$m_s = \frac{m}{2}(1 + \cos \beta)$$
$$m_r = \frac{m}{2}(1 - \cos \beta) \qquad (2.13)$$

This is a grossly over-simplified model as it assumes that the re-entrant jet is formed solely from the flyer plate, whereas at S there is a stagnation pressure which acts in both the parent and flyer plate materials. Consequently, the jet will be formed from the surfaces of both plates, as shown in Fig. 2.8, which is based on a schematic flow pattern proposed by Cowan and Holtzman (24). If sections DBA and EFG, in the flyer and parent plates respectively, are considered then layers BA and EF are removed in the jet and points F and B will be brought together as shown at D'B'F'G'.

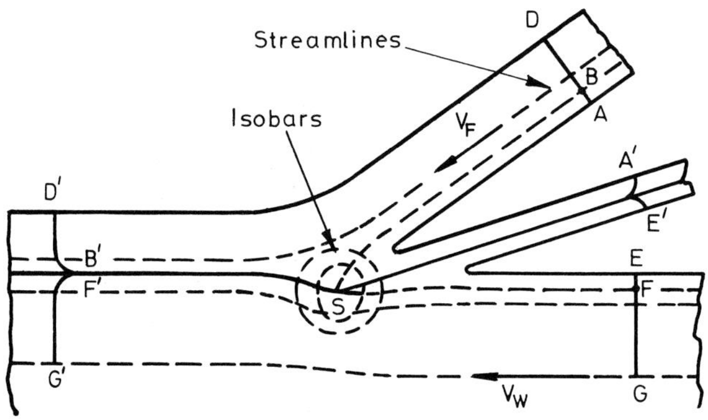

FIG. 2.8 Flow configuration in region of collision (24)

For the parallel set-up, which is used extensively for large plate cladding, it will be seen that Equations 2.10, 2.11 and 2.12 reduce to:

$$\sin \frac{\beta}{2} = \frac{V_P}{2V_D}$$

and (2.14)

$$V_W = V_F = \frac{V_P}{2 \sin \frac{\beta}{2}} = V_D$$

For an explosive with $V_D = 3$ km/s and for an explosive loading which gives a flyer plate velocity of 0.4 km/s, then for the parallel set-up $\beta = 15.3^{\circ}$, and the mass in the re-entrant jet will be 1.8% of the flyer plate mass. If the re-entrant jet is made up completely of flyer plate material, and the flyer plate is 10 mm thick, then the thickness of material removed from the flyer plate surface will be 0.18 mm. However, the jet consists of material from both the surfaces, and if it is assumed that the density of both the flyer and parent plate are the same, then it is perhaps not unreasonable to assume that the thickness of material removed from each surface is about 0.09 mm.

The existence of a well formed jet in conical shaped charges has been well established by flash X-ray photographs, and in wedge-shaped cutting charges its existence can be deduced from the observed penetration of the free jet into a target plate. In explosive welding the conditions are less severe, and in the parallel set-up in particular there has been some debate as to whether the jet really exists, or if it exists does it escape? However, Bergman et al (25) have studied this problem for a parallel set-up using a framing camera, and also by partially electro-plating copper on the surface of a nickel flyer plate, which was then welded to a steel parent plate. Their photographs establish that a jet in the form of a spray of metal particles is formed, and that in vacuum it travels at a velocity about twice the detonation velocity, which is what would be expected. In atmosphere the jet velocity is somewhat reduced due, no doubt, to the frictional effects of the compressed air. It was also noted that the temperature of the jet, as shown by its luminescence, was higher in air due to the adiabatic compression of the gas between the flyer and parent plate and the frictional heating. However, the presence of air apparently has no significant effect on the weld interface. From their experiments with a partially electroplated flyer plate they concluded that there was a substantial free jet which escapes, while a little of the jet which impinges on the two surfaces in front on the collision point is trapped. The existence of a free jet can also be deduced from the observation that metal surfaces close to an explosive welding operation exhibit evidence of a spray of material ejected from the system. The existence of a jet can also be deduced from the experiments using a gas gun which have been reported by McKee and Crossland (26), in which a flyer plate carried by a sabot is fired at an oblique target or parent plate. In these experiments the loss of mass was accurately measured, and a spray of ejected material could be observed on the target plate support anvil; under all conditions resulting in welding a loss of mass was observed.

It will be seen in this analysis of a very simple model of the explosive welding situation, that the surface layers which contain the oxides and adsorbed gases form the jet which is expelled from the system. If a sufficient thickness of material is removed from the surfaces then uncontaminated materials are brought together in the vicinity of the stagnation point, S in Fig. 2.8, where there is an intense pressure. It is difficult to conceive of better conditions for the formation of a solid state weld.

In the analysis considered so far the collision problem has been treated as one of classical hydrodynamics of an incompressible perfect fluid, and the role of shock waves within the flow has been ignored. This analysis predicts that a re-entrant jet will always exist, and that the re-entrant jet velocity approaches infinity as β is reduced to zero, while the thickness of the jet becomes vanishingly small. The analysis rests on the assumption of fluid-like behaviour under high velocity impact conditions, and this appears to be valid for the extreme conditions achieved in a hollow charge. Even under the less extreme conditions met in explosive welding the assumption leads to a reasonable physical insight, but nevertheless the pressure in the contact zone is not as great and the strength of the materials is not as insignificant as in the case of a hollow charge.

Walsh et al (27), Abrahamson (28) and Cowan and Holtzman (24) have used compressible flow theory to more fully explore the flow in hollow charges and in explosive welding. Walsh et al have mainly considered the symmetrical collision between two plates depicted in Figs. 2.2 and 2.3, for the case where the half angle, β, between them is small so that the velocity, V_2, of the liner relative to the observer moving with A is supersonic. They showed that such collisions can be divided into two categories, namely, jetless and jet forming collisions which are illustrated in Fig. 2.9. It was concluded that the only possible configuration for a jetless collision is that shown in Fig. 2.9a in which there is a single shock at A.

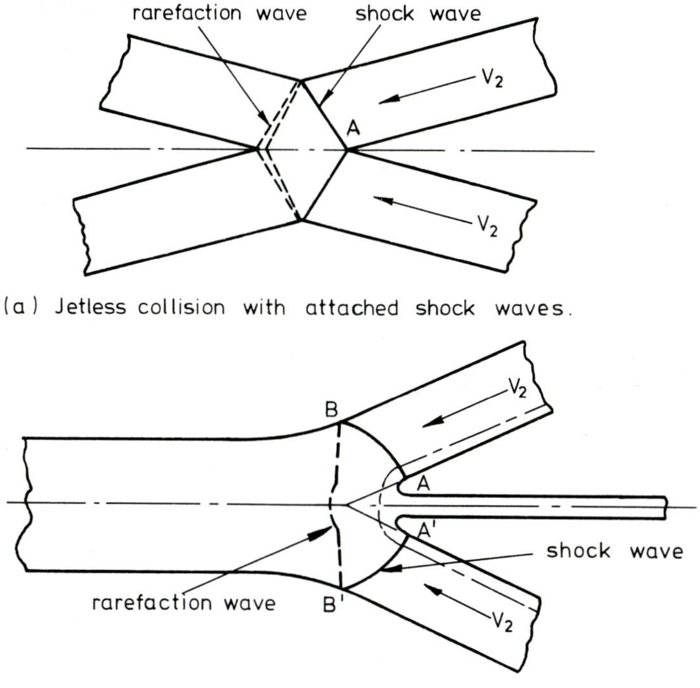

FIG. 2.9 Supersonic collision (27)

However, above a certain critical angle, β_C, which is dependent on V_2, the configuration with a single shock at A is no longer possible and a detached shock wave moves upstream as indicated in Fig. 2.9b. Jetting then occurs downstream from the detached shock waves due to the pressure acting on the unsupported flanks of the main jet. According to Cowan and Holtzman since the detached shock waves tend to deflect the flow away from the collision point, the collision angle is somewhat less, and consequently the mass in the jet may be expected to be less than for the subsonic case for the same angle β.

Cowan and Holtzman (24) have considered the asymmetrical case, involving the collision of plates of different materials and thicknesses, and travelling at different flow velocities relative to the collision region. In particular they have considered the case of explosive welding using the set-up depicted in Fig. 2.1, and including the parallel arrangement when $\alpha = 0$. For the parallel set-up they have calculated the critical angle for jetting for various metals and combination of metals, as a function of the detonation velocity, $V_D (= V_W = V_F)$, and the results are shown in Fig. 2.10 where jetting occurs to the left of the lines.

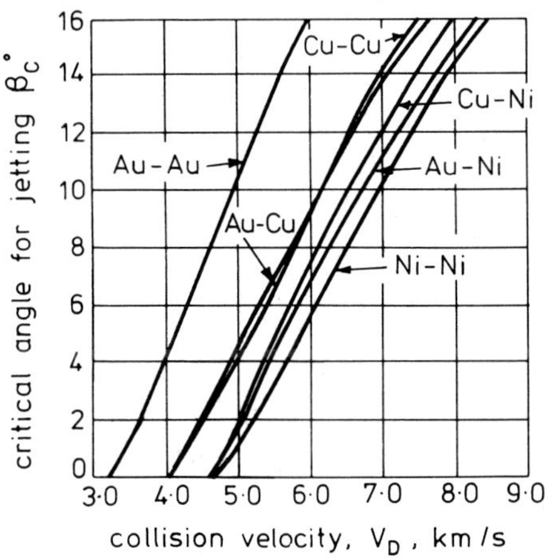

FIG. 2.10 Calculated values of β_c and V_D for parallel set-up. Jetting occurs to the left of lines (24)

Though the limiting conditions for jetting can be calculated little can be said about the flow pattern in asymmetrical collisions. If the two plates are the same metal but of different thickness, and they approach the observer at the collision point at the same velocity, then the stagnation pressure of both streams is the same. The flow in the region near the stagnation point is not dissimilar to the symmetric case, but the angle of the jet and the proportion of the two plates in the re-entrant and salient jets are uncertain. If the two plates are of different materials or approach the collision points at different velocities, then the stagnation pressures of the two streams are different. There are two possible flow configurations envisaged by Cowan and Holtzman, one in which only the stream with the lower stagnation pressure separates to form a re-entrant jet, and in the other a cusp is formed which permits both streams to divide to form a jet.

It has to be realised that the flow patterns for ideal fluids may not even be meaningful for real fluids, much more for metals, for at least two reasons. Firstly, the flow patterns envisaged contain regions where viscous effects could be especially important in modifying the fluid flow, such as the region of high relative velocity between the re-entrant jet, and the parent plate. Secondly, and perhaps more importantly the flow observed is frequently unstable and oscillations occur at the interface.

Though a perfectly acceptable weld can be produced with a straight interface, nevertheless under welding conditions frequently achieved in practice the weld interface shows well formed regular waves. It should be noted, however, that contrary to common belief a weld with a straight interface may be equally as strong as one exhibiting a wavy interface. Interfacial waves can be detected in explosive welds first reported by Carl (4), and Lavrentiev as reported by Deribas (5) also observed the wavy interface. Allen et al (12) also noted similar waves on the front of flat-nosed steel cylinders, fired at high velocities at oblique thin sheet metal targets of lead and copper. Even before Carl, however, waves had been observed on metal surfaces subject to oblique impact from solid shot, but its significance as a demonstration of the fluid-like properties exhibited was not appreciated.

The interfacial waves at the weld interface in explosive welds has stimulated much intellectual curiosity. A very useful experimental technique to obtain a better physical picture of the process was used by Bahrani et al (29) in their experimental investigation of the mechanism of wave generation. This consisted of electroplating a very thin layer of some trace metal on either or both the parent or flyer plate, so that when the plates are welded together it may be sectioned and the distribution of the trace material examined. In their experiments they used copper-plating to investigate the wave mechanism in mild steel to mild steel welds. An extension of this technique has been used by Williams (30) to study the welding of copper to copper, copper to lead and copper to titanium, in which he electroplated a large number of alternate layers of copper and nickel on to the copper plates. On sectioning after welding the deformation of the interfacial region can be studied in detail, rather like photographs of smoke filaments used in examining the flow of air round an obstacle. Figs. 2.11, 2.12 and 2.13 show micrographs of welds produced by Williams. The progressive development of interfacial waves from the initiation of welding for copper to copper is shown in Fig. 2.11. It will be seen that initially there is a straight interface with periodic oblique adiabatic shear lines, then smooth waves develop with no fore and aft regions of turbulence, and later waves can be observed with fore and aft regions of turbulence. Waves between materials of equal density appear to be nearly completely symmetrical, while with materials of different densities such as copper to lead and copper to titanium the waves are asymmetrical, and tend to have a single region of turbulence associated with each wave.

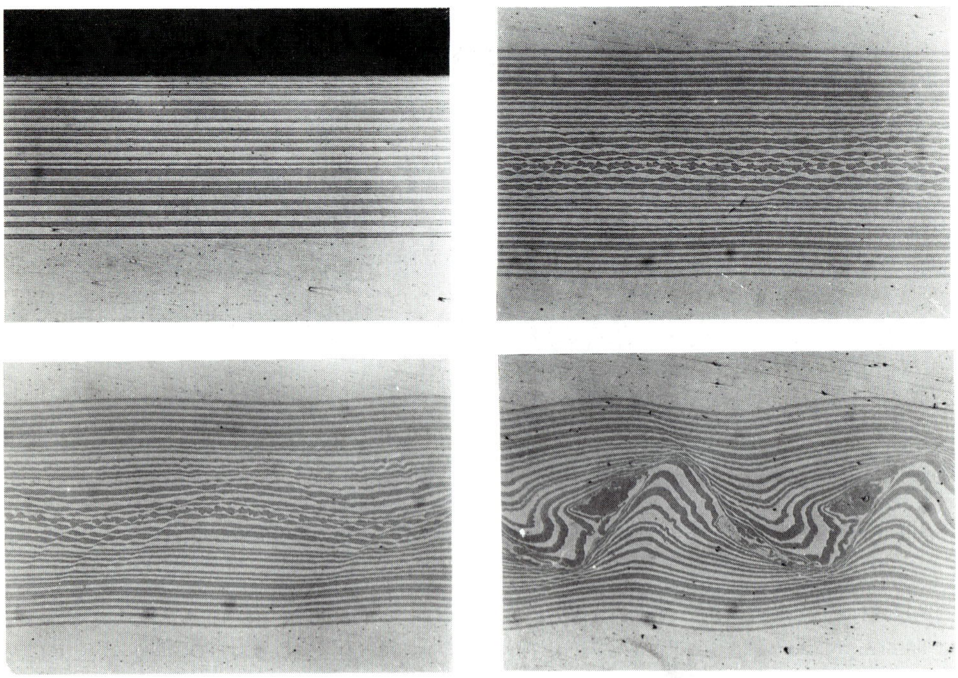

FIG. 2.11 Cu-Cu weld with multilayer parent and flyer plates, X 100. Direction of weld → (a) original plate materials, (b) start of weld, (c) start of waves, (d) fully developed waves

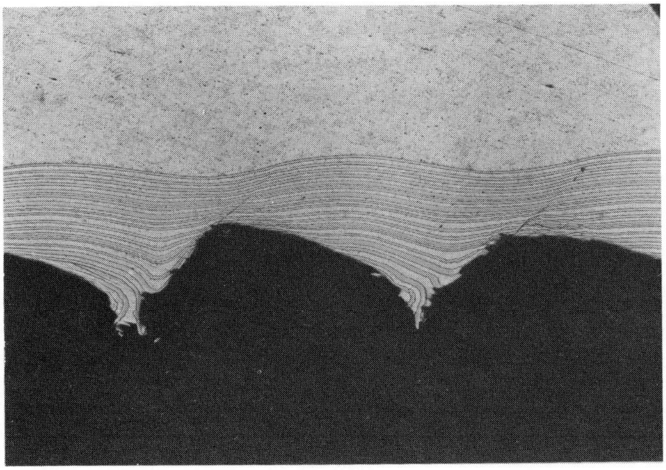

FIG. 2.12 Multilayer flyer plate welded to lead, X 60.
Direction of weld ←

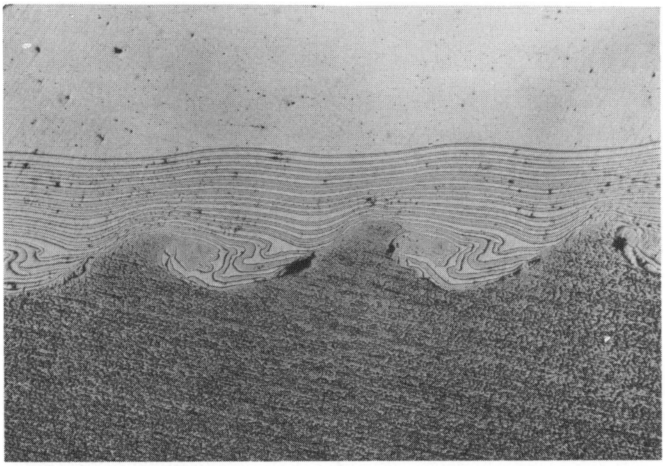

FIG. 2.13 Multilayer flyer plate welded to titanium, X 60.
Direction of weld ←

Various mechanisms have been proposed to explain the generation of waves in explosive welding, and it is convenient to divide them into two groups. Historically the first group can be classed as indentation mechanisms as typified by Abrahamson (28) and Bahrani et al (29), which provide a physical description of wave mechanisms. Abrahamson based his proposed mechanism on the waves observed in a tray of silicon putty, which was moved at a constant velocity under a fixed water jet, as he argued from the work of Allen, Mapes and Wilson (12) that the formation of waves was essentially a subsonic phenomenon. According to this mechanism the waves are created by the salient jet, and if this is so then it might be expected that the wavelength of the waves generated should be comparable to the thickness of the salient jet, which is very nearly the thickness of the flyer plate. In his experiments with a water jet impinging on moving silicon putty, the observed wavelength was typically three or four times the salient jet thickness. Bahrani et al (29) noted in their explosive welding experiments, that the wavelength of the waves generated were much less than the salient jet thickness. They also noted that the waves they were observing had been produced at the rate of 10^7 to 10^8 per second, and it was hardly conceivable that deformation waves could be laid down at this rate by the mechanism proposed by Abrahamson, which involves the entire thickness and momentum of the salient jet.

Bahrani et al (29) based their proposed mechanism on observations of explosive welding experiments, in which a nickel plated steel flyer plate was welded to a copper plated steel parent plate, which made it possible to detect what happened to the surface material of both plates. Fig. 2.14a shows the combination of the deformation of the parent plate, caused by the momentum of the flyer plate jet, coupled with the forward velocity of the parent plate relative to S, together with the shearing of the surface of the parent plate by the re-entrant jet. As a consequence a hump is formed ahead of the point of impact. This hump deflects the re-entrant jet upwards into the flyer plate jet as shown in Fig. 2.14b, and ultimately it completely blocks off the re-entrant jet as shown in Fig. 2.14c. Sometimes when the flyer plate jet impacts the crest of the wave it shears off part of the hump, giving rise to the tail illustrated. The trapped re-entrant jet forms a vortex at the back of the hump, in which high temperatures are likely to be created by the dissipation of the kinetic energy of the trapped jet, and this can result in phase changes and local melting. When the re-entrant jet is completely choked the stagnation point moves from the trough to the crest of the wave as shown in Fig. 2.14d, and the high pressure associated with the stag-

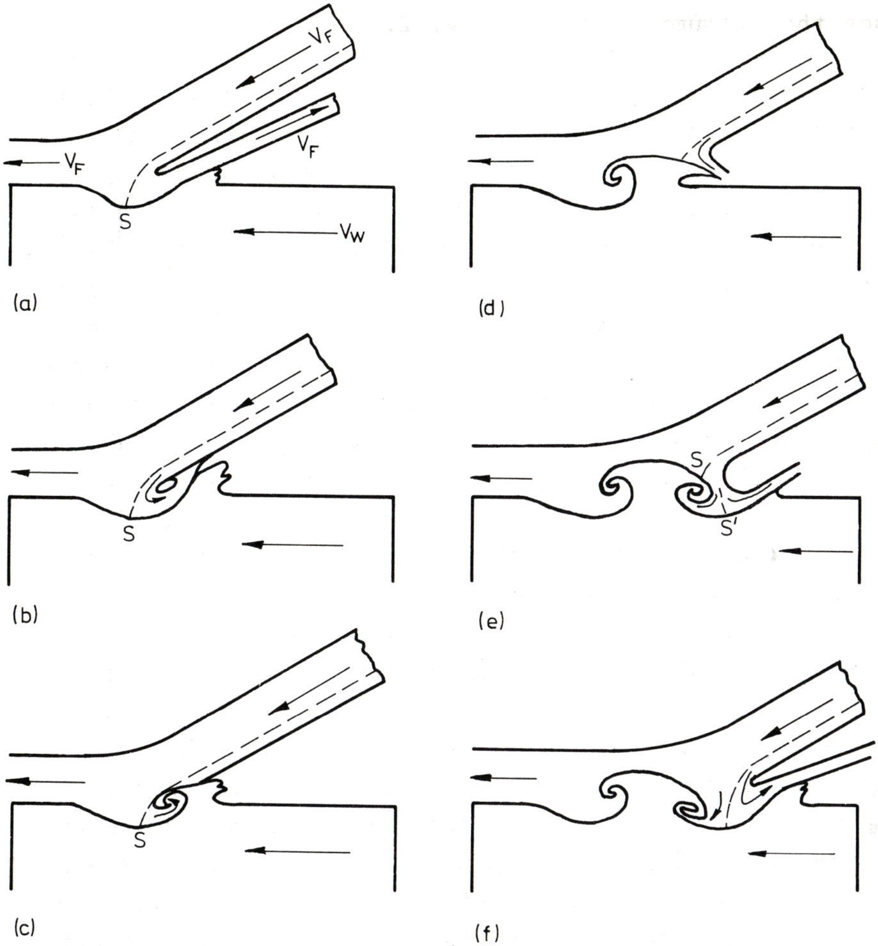

FIG. 2.14 Formation of waves according to Bahrani et al (29)

nation point will depress and elongate the hump so that a forward trunk is formed. As the hump continues to move downstream the stagnation point descends the forward slope of the hump as shown in Fig. 2.14e, and as a consequence the re-entrant jet will be increased due to the increased angle of inclination between the jet and the inclined side of the hump. At the same time the velocity of the re-entrant jet is reduced. As the re-entrant jet descends the forward slope of the hump a second stagnation point is formed at S', and part of the jet enters the cavity under the trunk causing another vortex. Besides creating a vortex under the trunk of the first wave the re-entrant jet sweeps the surface ahead of the first wave and it begins to produce the next hump as shown in Fig. 2.14f. The same mechanism will apply to the forming of the second and subsequent waves.

The mechanism proposed by Bahrani et al appears to be in reasonable accord with their experimental observations, and it is also supported by the experimental work of Wilson and Brunton (31). They were interested in the wavy interface produced on metal surfaces by high speed impact of liquid drops, such as in the rain erosion of high speed aircraft and in steam turbine blade erosion by wet steam. In order to investigate the wave mechanism they carried out an experiment in which impact between two sheets of liquid, one clear and one dyed, was observed with a high speed camera. They compared their observations with the mechanism proposed by several authors, and they concluded that the best agreement was with that of Bahrani et al. Kudinov and Bunyatyan (32) have also reported on similar experiments and the results look much the same as those of Wilson and Brunton.

The mechanism proposed by Bahrani et al has been modified by Lucas (33) to take account of the re-entrant jet being formed from both the flyer and parent plate. He notes that if the stagnation pressure is identical in both plates, then the velocity imparted to the material which enters the re-entrant jet will be greatest for the material with the lower density. As a consequence of the difference in velocity between the two component parts of the re-entrant jet, the jet will be deflected up or down depending on whether the density of the parent plate is less or greater than that of the flyer plate. Lucas uses this argument to explain why one or other of the vortices, frequently noted in welds between materials of equal density, disappear in welds between materials of significantly different density.

Though the mechanism proposed by Bahrani et al (29) seems in accord with experimental evidence, it is not adequate. Though it provides a good physical picture of the formation of the vortices before and after each wave, it does not provide an adequate explanation of the underlying wave mechanism. It does not provide an adequate physical framework to enable the periodicity of the interfacial waves to be related to the geometry of the set-up or to the properties of the materials.

The first group of papers concerned with the generation of waves are qualitative in their description. The second group of papers attempts to describe the generation of waves in more quantitative terms capable of predicting the conditions required for wave formation and the factors which control the size of waves. They are based on the recognition by many investigators of the similarity between the interfacial waves in explosive welding, and the waves observed in various fluid flow situations, such as those generated behind an obstacle in fluid flow or those formed at the interface of two fluid streams which have different velocities. However, just because there is a great similarity, does not mean that the cause of the two phenomena are necessarily the same. Even though the two metals in the region of the point of impact behave like fluids because of the very high instantaneous pressure, they behave as solids as soon as they leave this region, which is dissimilar to real fluids. The waves generated at the weld interface must be formed within close proximity of the collision point, whereas in real fluids the waves generated behind an obstacle or at the interface between two fluid streams grow over an appreciable distance. The great similarities between the two situations must be tempered with a realistic appraisal of the differences.

Cowan and Holtzman (24) were probably the first to recognise the obvious similarities between the interfacial waves in explosive welding and the fluid flow round an obstacle, in which regular formation of eddies is observed. Fig. 2.11 shows the interfacial waves formed between two copper plates, and for comparison Fig. 2.15 shows the Kármán vortex streets in the wake behind a circular cylinder in a stream of oil, reproduced from a paper by Homann (34). The visualization of the flow behind the cylinder was achieved by colouring matter released at the cylinder and carried downstream in the wake. It will be seen in Fig. 2.15 that the form of the wake depends on the Reynolds number. The Reynolds number, which is non-dimensional, is given by:

$$R = \frac{Vd\rho}{\mu} \qquad (2.15)$$

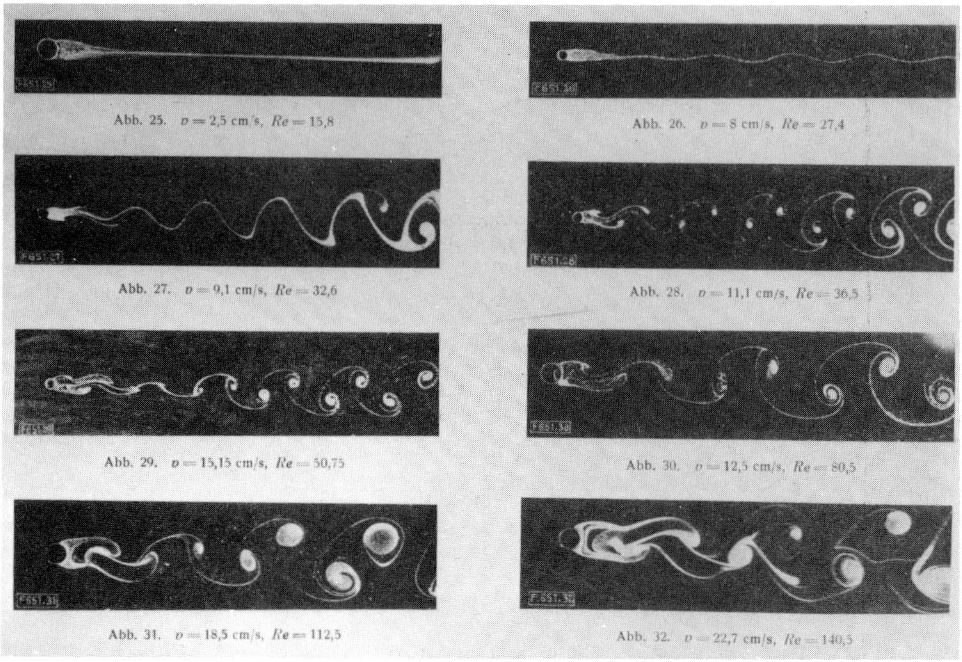

FIG. 2.15 Kármán vortex streets generated by a cylinder in fluid flow (34)

where V is the velocity of the fluid stream, d the diameter of the cylinder, ρ the density of the fluid, and μ the viscosity of the fluid. For R = 55 there are well developed smooth waves, but with increasing values of R the vortex patterns became increasingly irregular and turbulent in character.

The similarity between interfacial waves in explosive welding and the Kármán vortex streets in fluid flow passed an obstacle has been extensively studied by Cowan et al (35), Kowalick and Hay (36), Reid (37) and Reid and Sherif (38). According to Cowan et al (35) whether waves are formed or not, depends on the velocity of the flyer plate relative to the collision point, V_F, and the angle of obliquity, β. Fig. 2.16 shows the region in which waves are formed, and Cowan et al consider that this is made up of two boundaries. At high values of V_F, when the stagnation pressure is high, the flow can be considered as essentially Newtonian, and interfacial waves occur above a critical value of the Reynolds number. The dimension of the obstacle in the flow path is taken as the jet thickness. From Equation 2.13 the jet thickness is seen to be

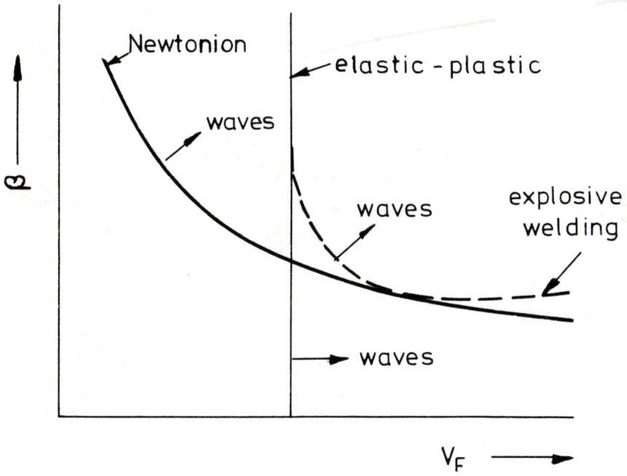

FIG. 2.16 Theoretical and practical boundaries for wave formation (35)

proportional to $\frac{1}{2}(1 - \cos \beta)$ or to $\sin^2 \frac{\beta}{2}$, which for small values of β implies that the jet thickness is proportional to β^2. Consequently, as ρ and μ are constant, the boundary line for a critical value of Reynolds number is defined by:

$$\beta^2 V_F = \text{constant} \tag{2.16}$$

At lower values of V_F the stagnation pressure is less, and the material may be more correctly considered as an ideal elastic plastic solid, and the Reynolds number must be defined in terms of the yield stress Y. From (2.15):

$$R = \frac{V_F d \rho}{\mu} = \frac{\frac{1}{2}\rho V_F{}^2}{\frac{\mu V_F}{2d}} \propto \frac{P_s}{Y} \tag{2.17}$$

where P_s is the stagnation pressure for an incompressible stream of density ρ travelling at velocity V_F. Thus the boundary line for a critical value of Reynolds number is defined by:

$$V_F = \text{constant} \tag{2.18}$$

Cowan et al established the conditions for the transition from smooth to wavy interfaces for various metal combinations. At the low velocity boundary they established that the transition velocity is largely independent of β. In this region they took the yield stress in Equation 2.17 as the diamond pyramid hardness. In metal cladding the flyer and parent plate materials are usually of different materials, and to allow for this they used the average values of density and hardness. So the Reynolds number becomes:

$$R = \frac{(\rho_F + \rho_P)V_F^2}{2(H_F + H_P)} \tag{2.19}$$

Using coherent units such as kg/m^3 for density, m/s for V_F and N/m^2 for hardness, they found that the value of Reynolds number at the transition was 10.6 with a relative standard deviation of 17.9%. From an examination of errors they concluded that the critical value of the Reynolds number is a constant, though there is a possibility that the value may vary for different classes of metals due to differences in strain rate dependence on strength properties, however, these effects are not large compared with the experimental error.

Cowan et al then went on to account for the size of waves, and in particular the experimentally observed fact that the size is strongly dependent on the size of the collision angle, and relatively independent of the collision velocity. They based their analysis on the analogy to the Kármán vortex streets behind an obstacle in fluid flow. If this is a really justifiable assumption must be debatable, as in the case of a near perfect fluid the stable wake width is achieved well downstream of the obstacle, whereas in explosive welding the waves are frozen before they could achieve a comparable condition of stability. Their analysis predicts that the wave amplitude is independent of the velocity V_F and dependent on β^2, which is in agreement with experimental evidence given in their paper. They also demonstrated that their experimental results led to a range in the unknown constants, involved in their analysis, which are plausible. Consequently, they concluded that the formation of the interfacial waves in explosive welding is basically similar to the development of Kármán vortex streets behind an obstacle, in particular a flat plate moving with a constant velocity through a fluid. Experiments with colliding flat liquid streams were also reported, in which a transition from smooth to oscillating behaviour was observed, which provided additional support for the analogy.

Kowalick and Hay (36) have also assumed that the waves in explosive welding are analogous to Kármán vortex streets, and as with Cowan et al (35) they assumed that the obstacle height is equal to the thickness of the re-entrant jet. Their physical picture of the wave generation process is shown in Fig. 2.17. In their experimental work they adopted the cylindrical plate technique used by Abrahamson (28) and Bahrani and Crossland (39) and shown in Fig. 2.18, which allows the effect of varying β to be examined in a single experiment. Using this

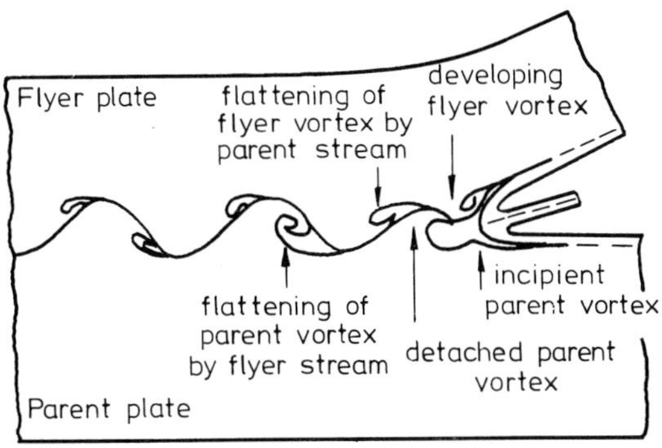

FIG. 2.17 Generation of periodic wave formation by vortex shedding (36)

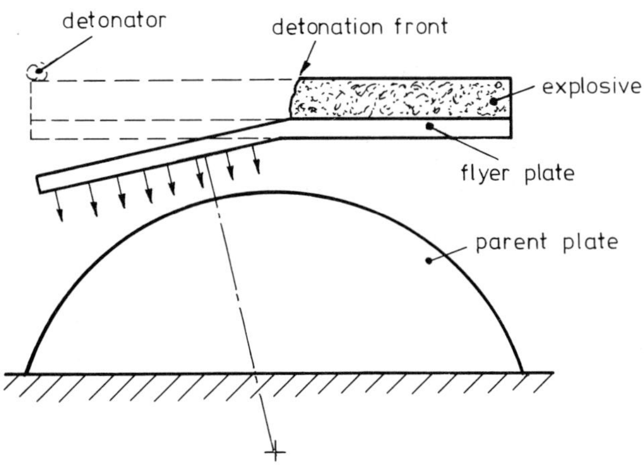

FIG. 2.18 Arrangement with semi-cylindrical parent plate (36)

technique for steel to steel welds, and taking a value of viscosity of 2 x 10^4 poise, they found the transition from laminar to wavy flow occurred for values of Reynolds number lying between 2 and 3.2, and wave formation continued up to a Reynolds number of about 10. The assumed value of viscosity was that given by Harlow and Pracht (40), and Kowalick and Hay realised that it may be in considerable error, but it is reasonably in agreement with the value of 10^4-10^5 poise deduced by Klein (41).

Reid and Sherif (38) have also based their work on the Kármán vortex street analogy. Considering a symmetric explosive welding situation, such as is depicted in Fig. 2.5, they assumed that the obstacle in the flow path was cylindrical with a radius R_S, as shown in Fig. 2.19. Using this obstacle they showed that the wavelength was approximately $4t\beta$, which was in reasonable agreement with experimental data. They did not, however, consider the limiting conditions between laminar and wavy flow.

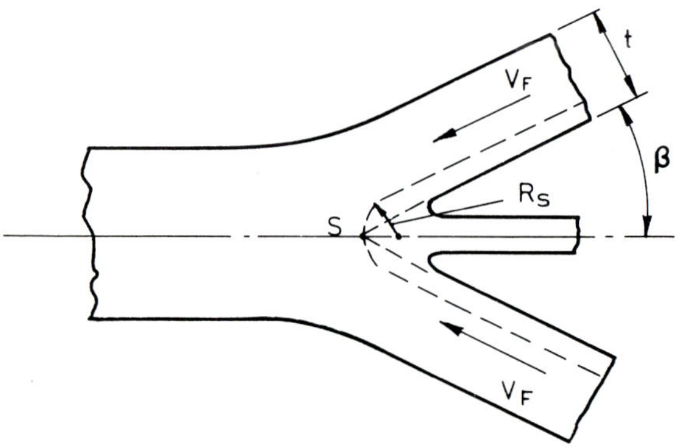

FIG. 2.19 Cylindrical obstacle assumed by Reid and Sherif (38)

An alternative fluid mechanics analogy is that the waves at the interface in explosive welding are similar to a fluid interface exhibiting Helmholtz instability. Helmholtz instability describes the behaviour at a fluid interface, across which there is a discontinuity in fluid velocity. Hunt (42) was probably the first to suggest this analogy, and he assumed that the waves were generated ahead of the contact point, by the relative velocity between the re-entrant jet and the parent plate. He assumed that the re-entrant jet was formed from the flyer plate, and that it kept in contact with the parent plate as shown in Fig. 2.7. However, this gives a very short length of time for the waves to grow under the relative velocity between the re-entrant jet and the parent plate, as the zone around the contact point in which the pressure is sufficiently high for the metals to behave as fluids is very small, and outside this zone the waves effectively become frozen. Hunt assumes that the re-entrant jet consists purely of flyer plate material, which has a velocity relative to the parent plate, whereas there is little doubt that the re-entrant jet normally consists of the surface layers of both the flyer and parent plates, which is radically different to the model he has assumed. If waves are generated by Helmholtz instability, then in a symmetrical weld such as that depicted in Fig. 2.5 there should be no waves, as there is no relative velocity between the two components of the re-entrant jet. However, both Cowan and Holtzman (24) and Onzawa and Ishii (44) have observed interfacial waves in symmetrical welds, and this must cast doubt on the validity of the analogy proposed by Hunt.

Onzawa and Ishii (44) and Klein (41) (45) have also assumed that the waves are created by the discontinuity of velocity across the interface. However, Klein considers that the waves occur at the interface between the salient jet and parent plate, not the re-entrant jet and parent plate, which would suggest that in symmetrical welds waves would not be generated. Though Keller (46) did not find waves in symmetrical welds other experimenters have observed waves, which suggests that the analogy assumed by Klein is not valid. Klein also deduced the viscosity from his explosive welding experiments, and from metallurgical examination he deduced that the interface temperature was in the region 1300-1500°C. He compared this data with viscosity data deduced from high speed hot punching tests, and data deduced from hot rolling tests at lower strain rates, and he considered that all the data lay remarkably on one curve. However, the validity of the assumption made in collecting this data is open to question, and that the data lies reasonably on one curve may be fortuitous.

Godunov et al (47) considered the hydrodynamic effects in colliding solids, and they describe experiments designed to study the mechanism of initiation of waves. They noted that waves appeared at a certain distance from the start of collision between the plates, and they then grew gradually until they reached a steady state after a few oscillations. They concluded that the disturbance which caused the initiation of waves, came from the rarefaction wave originating from the start of collision, when it coincided with the current point of contact, as illustrated in Fig. 2.20. The distance, l, from

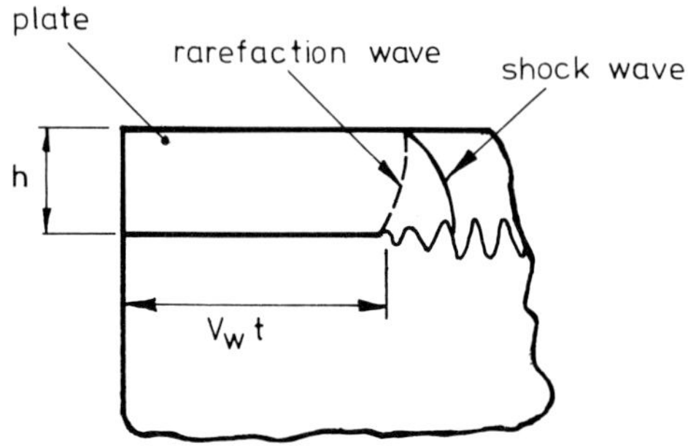

FIG. 2.20 Initiation of waves by rarefaction waves (47)

the start of collision to the point where the current contact point coincides with the rarefaction wave can, with the aid of linear acoustics, be shown to be:

$$l = \frac{2hV_W}{\sqrt{C_o^2 - V_W^2}} \tag{2.20}$$

where h is the thickness of the flyer plate, V_W the velocity of weld propagation, and C_o is the velocity of sound in the flyer plate material. They claimed that their experiments showed that waves appeared at approximately distance l from the start of collision. Estimates of the width of the rarefaction wave showed that it was of the order of magnitude of the wavelength of the waves generated. They also carried out tests in which a small step was machined on the parent plate surface, as shown in Fig. 2.21, and it was observed that if the size of the step was within the size of the steady state waves which would be generated, then steady state waves were generated after the step independently of the distance of the step from the start of the collision. This suggests that some form of disturbance is required to initiate waves. If the step was much smaller or larger than the steady state size of waves, it was observed that waves were not generated at the step. McKee and Crossland (26), however, noted some anomalous behaviour with a step in the parent plate surface in their gas

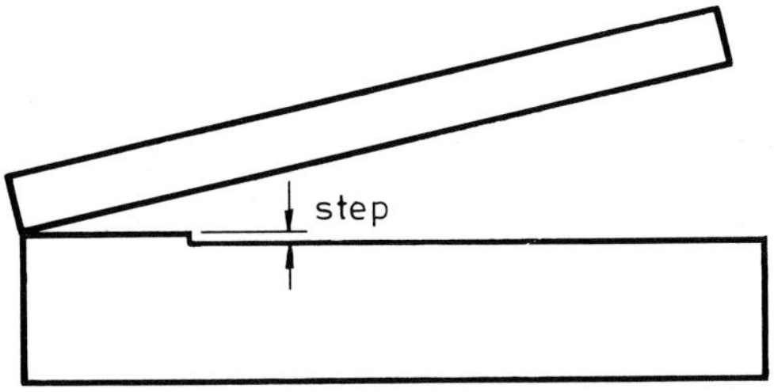

FIG. 2.21 Step in parent plate to initiate waves (47)

gun experiments. At low velocity the waves initiated before the step were maintained by the step, and in other cases the waves before the step were completely damped out by the step and there was no indication of re-initiation. At high velocity the step appeared to have a negligible effect on the waves.

Deribas (48) dismissed the Kármán vortex street analogy, on the grounds that waves could be observed in explosive welding which exhibited no turbulent vortices, though as can be seen in Fig. 2.15 this can also be true of classical Kármán vortices in liquids. From the work of Godunov et al he concluded that the interfacial waves in explosive welding are caused by a self-excited vibration excited by a rigid initiator, though there is no indication of what this might be. He tentatively suggests that this can in no way be considered as a finalized theory, and that the problem of an adequate mathematical model of the process has not yet been resolved.

El-Sobky and Blazynski (49) have also suggested a tentative stress wave mechanism of wave formation which is shown schematically in Fig. 2.22. They consider that a single com-

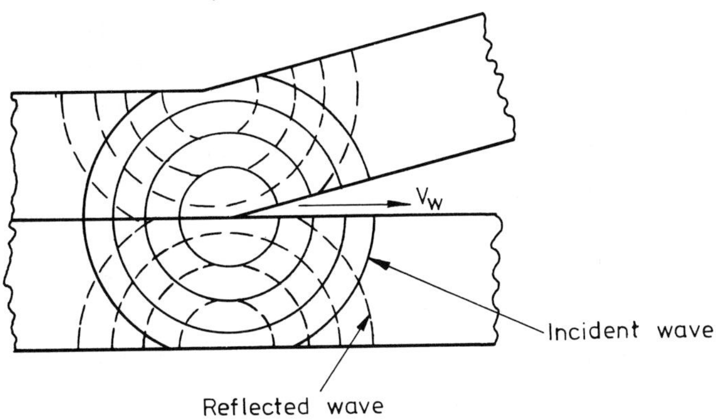

FIG. 2.22 Stress wave mechanism of surface wave formation (49)

pression wave represented by a full circle is generated at the collision point, while the dashed circles represent the successive reflection from both the flyer and parent plate. These reflected waves meet the surface periodically at points ahead of and behind the collision point constituting a continuous source of surface instability as the collision moves forward. If these stress waves are of sufficient intensity to produce surface deformation the collision would take place between two surfaces, which were already wavy. In such a case if these waves were in phase a clean wavy surface would be produced, and if they were out of phase then interaction would produce wave flattening and melt pockets which are sometimes observed. On the other hand if the stress intensity of the waves was not high enough to produce surface deformation, then only elastic surface waves would be created. The authors attributed the existence of continuous melt layers or small vortices with or without molten pockets to the interaction of the jet with this wave mechanism. This is a very similar mechanism to that proposed by Chadwick et al (50).

As can be seen the problem of adequately describing the mechanism of wave initiation is still by no means resolved. It must be realised that there is no inherent reason why different mechanisms should not operate for different material combinations and different values of the other variables. In many ways the Kármán vortex stress analogy seems to provide a likely explanation, but the problem is why does it require a finite distance from the point of first impact before waves are initiated, and why can a step in the parent plate act as an initiator? Even if the Kármán vortex street analogy is adequate there are still formidable problems in predicting the amplitudes and wavelengths of the waves generated. One of the essential problems is in measuring and describing the properties of the metals being welded under the extreme conditions experienced in the contact region. Fortunately, though the problem of the mechanism of wave initiation is an intriguing one, nevertheless its solution is not essential to the practical application of explosive welding.

3

EXPLOSIVES AND THEIR PROPERTIES

There are many authoritative books written on chemical explosives such as Taylor (51), Cook (52), Rinehart and Pearson (53), Fordham (54) and Johansson and Persson (55), but it is necessary for the sake of continuity to include a brief resumé on explosives, and especially to discuss those explosives of particular relevance to explosive welding.

Chemical explosives are substances which can undergo rapid chemical reaction during which large quantities of heat and gaseous products are evolved, and which can exert a large dynamic pressure on the surroundings. Explosives can be considered to be of two types - burning or deflagrating explosives such as gunpowder or cordite and detonating or high explosives such as dynamite or trinitrotoluene (TNT). Detonating or high explosives can be further divided into initiating or primary explosives, which are mainly used in detonators, and secondary explosives. It should be noted, however, that this division between deflagrating and high explosives is not clear cut, as some explosives under certain conditions will burn while under other conditions they will detonate.

Deflagrating explosives cannot generate the very high contact pressure, when in contact with a metal plate, which is required to achieve the acceleration needed if the flyer plate is to reach the high velocity necessary for explosive welding. It is difficult, or probably impossible, to confine the explosive in order to increase the pressure generated. With high explosives very large contact pressures can be generated without the need for confinement, consequently high explosives are universally used in explosive welding. However, in explosive cladding of large plates it will be noted that it is desirable or necessary to use an explosive with a detonation velocity less than the sonic velocity in the metals being welded, which implies a detonation velocity less than 4 km/s or even less for some metal combinations. This eliminates quite a few explosives, and in particular sheet explosives, which otherwise might be more convenient to handle than powder explosives.

Another disadvantage of using high detonation velocity explosives, say with a detonation velocity of 7 km/s, is that the magnitude of the pressure pulse is so large that if it is reflected from the free surface of the flyer plate while it is still in flight, then the magnitude of the tension wave may be sufficient to cause a tensile failure, or what is known as spalling.

In the case of low intensity sound waves, where the pressure and density remain sensibly constant, a sinusoidal wave will be transmitted indefinitely as a sinusoidal wave. However, with an intense half sinusoidal wave the pressure and density cannot be assumed constant, and as the pressure is increased the compressibility decreases, so that the velocity of propagation of a disturbance increases. Consequently, the velocity of propagation of the peak of the wave is greater than that at the beginning of the wave, where the pressure is lower. As a result the peak will overtake the beginning of the wave, and the wave form becomes a sharp fronted discontinuity or shock wave. Without additional energy a shock wave will, due to dissipation of energy, gradually reduce to an ordinary sound wave.

In a detonation wave a sharp pressure discontinuity travels through the explosive at high velocity, and the chemical reaction occurs in the close vicinity of the detonation front in an extremely short interval of time. It is the energy liberated in the detonation front which maintains the detonation wave despite dissipation of energy. Detonation velocities vary from about 1-9 km/s, depending on the chemical and physical nature of the explosive. Normally the detonation wave travels at a constant velocity. As the reaction time involved is so short, little or no expansion of the high temperature gaseous products is possible in the reaction zone, and as a consequence these gases occupy about the same volume as the explosive immediately before the reaction. This is the main reason why the pressure behind the detonation front is so high. With military explosives the pressure generated behind the detonation front is of the order of 19-35 GPa, while commercial explosives give a lower pressure.

In the steady state a detonation wave can be considered

as a sharp discontinuity travelling through the explosive at a velocity, V_D, as shown in Fig. 3.1a. Ahead of the detona-

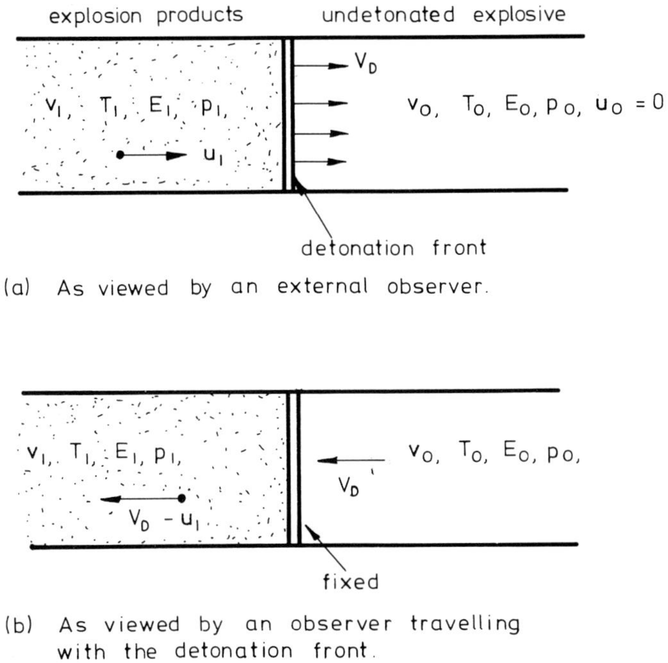

FIG. 3.1 Idealised detonation wave

tion front the state of the explosive can be specified by its specific volume, v_o, or the corresponding density, ρ_o, together with the specific internal energy E_o, temperature T_o and pressure p_o, while behind the front the corresponding values are v_1 or ρ_1, E_1, T_1 and p_1 and additionally there is a streaming or particle velocity u_1. Fig. 3.1b shows the situation viewed by an observer travelling with the detonation front. Applying conservation of mass, momentum and energy to the system shown in Fig. 3.1b gives the following equations:

$$\frac{V_D}{v_o} = \frac{V_D - u_1}{v_1} \tag{3.1}$$

$$\frac{V_D^2}{v_o} + p_o = \frac{(V_D - u_1)^2}{v_1} + p_1 \tag{3.2}$$

$$E_o + \tfrac{1}{2}V_D^2 + p_o v_o = E_1 + \tfrac{1}{2}(V_D - u_1)^2 + p_1 v_1 \tag{3.3}$$

It should be noted that E_o includes the internal energy, E_{oo}, and the chemical energy released, Q.

From Equations 3.1, 3.2 and 3.3 it can be shown that:

$$p_1 - p_o = \frac{V_D{}^2}{v_o{}^2}(v_o - v_1) \text{ or } \frac{V_D u_1}{v_o} \qquad (3.4)$$

$$V_D = v_o\sqrt{\frac{(p_1 - p_o)}{(v_o - v_1)}} \qquad (3.5)$$

$$u_1 = (v_o - v_1)\sqrt{\frac{(p_1 - p_o)}{(v_o - v_1)}} \qquad (3.6)$$

$$E_1 - E_o = \tfrac{1}{2}(p_1 + p_o)(v_o - v_1) \qquad (3.7)$$

From Equation 3.4 it must be concluded that the specific volume behind the detonation front is less than that ahead of the front. Consequently u_1, the streaming or particle velocity, is in the same direction as the detonation velocity.

It is possible to determine all values of p_1 and v_1, which satisfy Equation 3.7, if the equation of state of the products of the explosive reaction is known. This curve, which is known as the Rankine-Hugoniot curve, is shown as AB in Fig. 3.2. As v_1 is less than v_o, then the point which coordinates p_1, v_1, must lie to the left of the point representing the undetonated explosive, p_o v_o. The question which has to be answered is where does the point p_1 v_1 lie on the Rankine-Hugoniot curve?

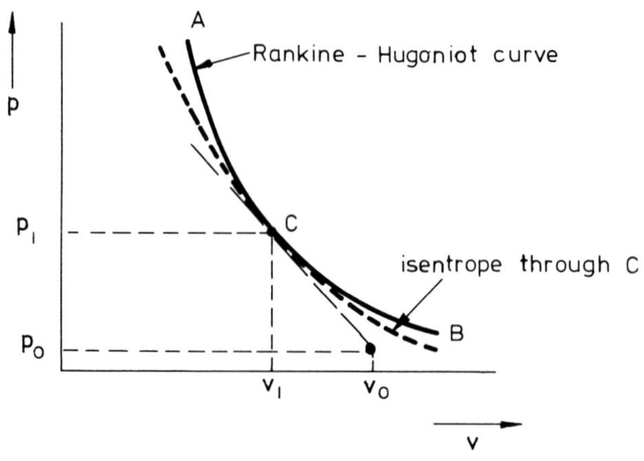

FIG. 3.2 Rankine-Hugoniet curve

The equilibrium point for the products behind the detonation front is the point of tangency between a straight line drawn from the point p_o v_o, which represents the undetonated explosive, and the Rankine-Hugoniot curve. This is the point C shown in Fig. 3.2 which is known as the Chapman-Jouguet point, and its justification is given by Taylor (50). It is then possible to determine the detonation and streaming velocities from Equations 3.5 and 3.6.

In the steady state the propagation velocity of a pressure disturbance immediately behind the detonation front, V_s, will be equal to the difference between the detonation velocity and the streaming velocity or:

$$V_s = V_D - u_1 \tag{3.8}$$

and from Equation 3.1:

$$\frac{V_s}{V_D} = \frac{v_1}{v_o} \tag{3.9}$$

However, the sonic velocity, V_s, can also be determined from thermodynamics which gives:

$$-\rho^2 V_s^2 = \left(\frac{\partial p}{\partial v}\right)_s$$

or,

$$-\frac{V_s^2}{v^2} = \left(\frac{\partial p}{\partial v}\right)_s \tag{3.10}$$

where $\left(\frac{\partial p}{\partial v}\right)_s$ is the change of pressure with specific volume at constant entropy. So that at the Chapman-Jouguet point:

$$\frac{V_s^2}{v_1^2} = -\left(\frac{\partial p}{\partial v}\right)_{s_1}$$

which from Equation 3.9 can be put in the form:

$$\frac{V_D^2}{v_o^2} = \left(\frac{\partial p}{\partial v}\right)_{s_1} \tag{3.11}$$

From Equation 3.5 it can be seen that:

$$\frac{V_D^2}{v_o^2} = \frac{p_1 - p_o}{v_o - v_1} \tag{3.12}$$

Comparing 3.11 and 3.12 it will be seen that the tangent to the Rankine-Hugoniot curve, drawn from the point p_o v_o, must also be the tangent to the isentrope drawn through the Chapman-Jouguet point. It must be concluded that the Rankine-Hugoniot curve and the isentrope must be coincident, and with a common slope, at the Chapman-Jouguet point as illustrated in Fig. 3.2.

The difficulty in applying the hydrodynamic theory of dedonation is in the equation of state, and the chemical equilibria for the detonation products. The equations of state for the various products of combustion have been extensively studied, for instance, see Taylor (50), but though there is some experimental confirmatory data, it is only up to pressures which are appreciably less than those experienced behind the detonation front.

Even if the equations of state are established, the detonation velocity calculated will be for a plane detonation wave, whereas for explosive charges of limited dimensions the detonation front will be curved due to edge effects, which will affect the detonation velocity. Additionally, in charges of limited size there may be a reduction of the driving pressure by lateral expansion, or by incomplete combustion. Whatever the cause, it is found with the lower detonation velocity explosives, in particular, that there is a significant reduction in detonation velocity with thickness or diameter of the explosive charge. There is also a minimum thickness or diameter below which detonation cannot be assured. Such effects cannot be adequately described theoretically, and consequently there is a need to establish such data experimentally.

As can be seen from Equation 3.5 the detonation velocity will be dependent on the specific volume or density of the explosive, and over a limited range, such as $\rho_o = 1000$ to $\rho_o = 1600$ kg/m^3, a good approximation is that detonation velocity is proportional to the density or:

$$V_D = A\rho_o \text{ or } \frac{A}{v_o} \qquad (3.13)$$

where the constant A depends on the explosive. For instance, for TNT when $\rho_o = 1640$ kg/m^3 the measured detonation velocity is 6.95 km/s, which gives 4.24 m^4/kg.s for constant A. From the equation of state of the gases behind the detonation front $\rho_1 = 2160\ kg/m^3$. Substituting this data into Equations 3.4 and 3.6 gives a detonation pressure of 19.1 GPa and a streaming or particle velocity of 1.673 km/s, which is about a quarter of the detonation velocity.

There are several methods available for measuring the detonation velocity in condensed explosives, which involve various degrees of sophistication. Perhaps the simplest method is that due to Dautriche (56), which is still extensively used for routine determinations by unskilled personnel. Essentially the detonation velocity is measured by comparison with a known standard. Fig. 3.3 shows the basic set-up for the Dautriche method; a length of detonating fuse of known detonation velocity, V_d, is connected to two points in the explosive charge, which are separated by a distance ℓ_1. The mid-point of the length of detonating fuse is known, and it is placed on a thin lead or soft aluminium plate and taped to it. The detonation waves travelling from each end of the fuse meet at a point distance ℓ_2 from the mid-point of the fuse,

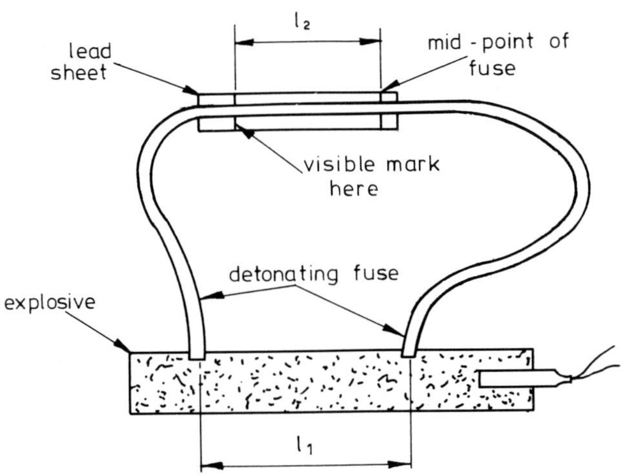

FIG. 3.3 Determination of detonation velocity by Dautriche method (56)

and a distinct mark is made on the plate at this point due to the intensification of pressure when the two detonation waves meet. Knowing V_d, ℓ_1 and ℓ_2 the detonation velocity in the explosive can be determined from:

$$V_D = \frac{V_d \ell_1}{2\ell_2} \tag{3.14}$$

The detonation velocity of standard detonating fuse is reasonably reproducible, but if greater accuracy is needed then the detonation velocity of a sample of fuse cut from the same reel may be checked by a pin insertion method or one of the other techniques available.

The Dautriche method is still commonly used as it is cheap and simple, even if crude, and it can be used on explosive sites without the problem of remote positioning of instrumentation. If V_d is 7 km/s and ℓ_1 is 150 mm then for V_D = 4 km/s the distance ℓ_2 will be 131.3 mm, which can probably be measured within 2%.

If a sophisticated recording system is available for recording electrical pulses, then there are various electrical methods for measurement of detonation velocity. In the explosive welding process when the flyer and parent plates are initially parallel, then the collision point velocity is equal to the detonation velocity. This is used as the basis of the pin contactor method shown in Fig. 3.4, in which the pins are

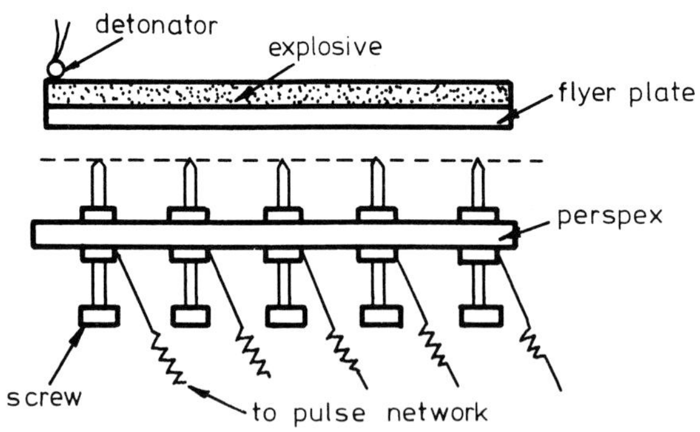

FIG. 3.4 Determination of detonation velocity using pin contactors and parallel plate set-up

adjusted to the same distance below the flyer plate and they are spaced 25 mm apart. The pins are connected to a pulse network, such as that described by Minshall (57), which produces a series of positive and negative pulses and these are recorded on a recording oscillograph. Yet another method is to use the pin insertion method shown in Fig. 3.5, in which pairs of pins

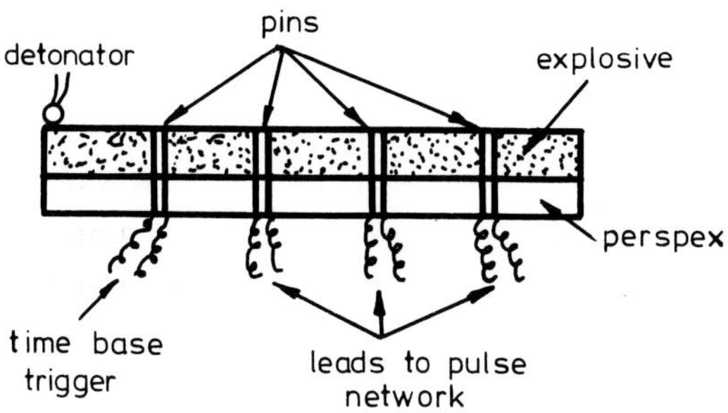

FIG. 3.5 Determination of detonation velocity by pin insertion set-up

are placed close together at various distances from the point of detonation, and they are connected to a pulse network and recording oscillograph as in the previous method. The shock wave, which is ionised, shorts the pairs of pins as it reaches them, and this triggers the pulse network.

Photographic methods can also be employed to measure the detonation velocity, and this has the advantage that a continuous record can be obtained of the velocity of detonation. The technique usually employed is to take a streak photograph, in which a slit is placed in front of the explosive and perpendicular to the detonation front. This slit is filmed by a rotating mirror camera which throws an image of the slit on to a fixed film, and it is so arranged that the movement of the image is at right angles to the direction of detonation. Consequently, the resulting photograph consists of a line inclined to the direction of movement of the image, and the slope of this line at any particular instant is related to the detonation velocity and the velocity of the image or the rotational speed of the mirror. With modern developments the photographic system would be replaced by an image converter tube, so that the velocity of the image can be measured electronically.

Besides detonation velocity an equally important property is the strength or power of an explosive, which is the area under the pressure/time record. However, to calculate this or measure it directly would be difficult or impossible. Two practical tests are widely used to obtain a measure of the power of an explosive. The first of these is the ballistic mortar described by Marshall (58), and standardised by an international committee in 1962. The ballistic mortar consists of a mortar barrel supported at the end of a pendulum; an explosive charge of 10 g is used in the mortar to fire a standard shot, and the angle of recoil is measured. It is general practice to use a standard explosive as a reference material, in Great Britain this is blasting gelatine and in the U.S.A. it is usually TNT. The results are expressed as a percentage of the strength of the reference explosive. An alternative test is the Trauzl test which is fully described by Marshall, in which the explosive is placed in a cylindrical hole in a lead block, the remainder of the hole is filled with sand. The expansion of the hole when the charge is detonated is found by measuring the volume, and the increase in volume is an indication of the power.

Another important property when it comes to handling an explosive is its sensitivity, or the ease with which it can be initiated by a normal detonator. Sensitivity is measured by using detonators of increasing strength, to find what strength is needed to initiate detonation in the explosive under test. Many secondary explosives are so insensitive that a detonator is insufficient to initiate detonation, or in the case of a few explosives it may initiate the lower of two apparently stable detonations, for instance see Taylor (50), which gives a lower pressure and inefficient energy release. To ensure the rapid initiation of stable detonation or the initiation of the higher velocity detonation, it is necessary to use a small booster charge between the detonator and the main charge. The booster is a secondary explosive in which detonation is readily initiated by a detonator. The mass of the booster charge to initiate stable detonation can be used as a measure of sensitivity, though no standard test has been proposed or adopted.

Though explosives must be sufficiently sensitive to be detonated, preferably by a detonator, they must be sufficiently insensitive to friction and impact to render them safe to handle. The sensitivity of friction can be determined by loading the explosive against a rotating drum or disc, and measuring the load and speed to cause detonation of the explosive. Another test is to measure the sensitivity of an explosive to a glancing blow, using the torpedo test illustrated in Fig. 3.6, in which the height of release to cause detonation is determined. Sensitivity to impact is measured by some form of drop rig such as that shown in Fig. 3.7, in which the height of fall to initiate detonation is measured.

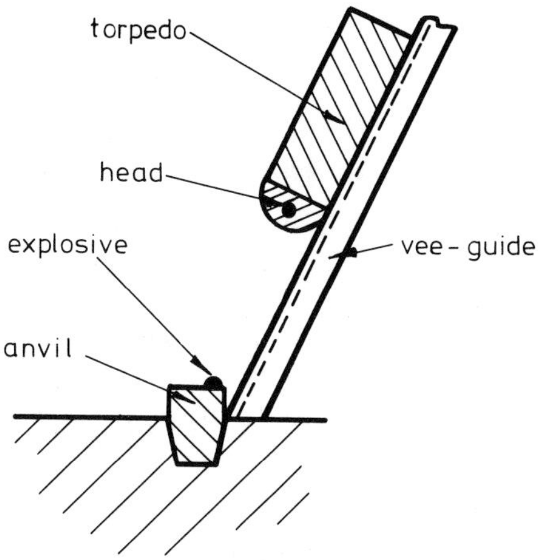

FIG. 3.6 Torpedo test (54)

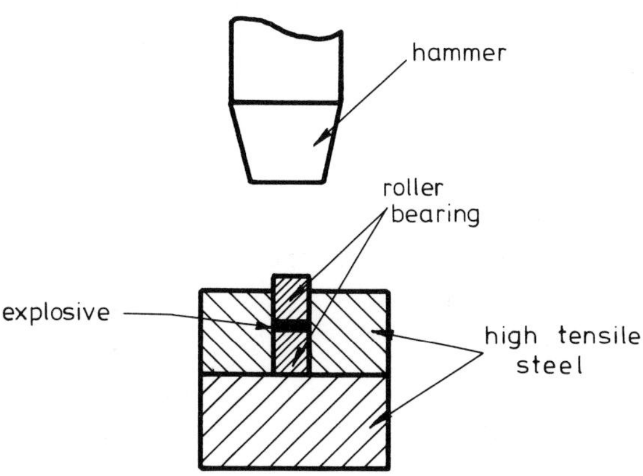

FIG. 3.7 Fall hammer test (54)

Explosives have been mainly developed for quarrying and mining operations, and for military purposes, but not to satisfy the needs of explosive welding. As the demand for explosives for welding operations is a very small fraction of that required for other purposes, there has been little incentive to develop explosives specifically designed for welding operations. Consequently, in explosive welding use has had to be made of available explosive technology, except by the main commercial organisations involved in large plate cladding, who are mostly off-shoots of manufacturers of explosives. Probably these organisations have the facility to control the detonation velocity, within limits, and to ensure the homogeneity of the explosives they use, but they regard such information as a trade secret.

As will be appreciated from the discussion on welding parameters in Chapter 5, there is a serious restriction on the explosives which can be used, particularly for large plate cladding, where the detonation velocity should be less than the sonic velocity in the materials beingwelded. Also, in order to achieve a reasonable value of the dynamic angle of obliquity, β, without excessive kinetic energy, it is necessary to keep the detonation velocity low. The following information on explosives will mainly be restricted to those of value in explosive welding processes.

Table 3.1 gives the primary explosives which are mainly used in the priming charge of detonators, which will be described later. Table 3.2 gives a selection of single substance explosives, some of which are used for the base charge in detonators or as booster charges to initiate detonation in low sensitivity explosives, while others are components of explosive mixtures, some of which are listed in Table 3.3.

TABLE 3.1

Primary Explosives

Name	Chemical Composition
Lead trinitroresorcinate (Lead styphnate)	$Pb\ C_6\ HO_2\ (NO_2)_3$
Lead azide	$Pb\ N_6$
Mercury fulminate	$Hg\ C_2\ N_2\ O_2$

TABLE 3.2

A Selection of Single-Substance Secondary Explosives

Name	Chemical Composition
Nitroglycerine (NG)	$C_3 H_5 (NO_3)_3$
Nitrocellulose (CC)	$C_{24} H_{40-x} O_{20-x} (NO_3)_x$
Trinitrotoluene (TNT)	$C_7 H_5 (NO_2)_3$
Pentaerythritol tetranitrate (PETN)	$C_5 H_8 (NO_3)_4$
Cyclotrimethylene trinitramine (RDX)	$C_3 H_6 N_3 (NO_2)_3$
Trinitrophenylmethynitramine (Tetryl)	$C_7 H_5 N (NO_2)_4$
Nitroguanadine	$C H_5 N_3 NO_2$

It is difficult to give data, such as relative power, sensitiveness, detonation velocity, etc., for the explosives noted in Tables 3.1, 3.2, and 3.3, as such data does not seem to be readily available or easily relateable to particular explosive compositions. Additionally, detonation velocity depends on the dimensions of the charge, its density and the degree of confinement. However, an attempt has been made in Table 3.4 to give reasonably meaningful data, as an indication of the properties of some of the explosives so far mentioned.

Most of the explosives listed in Table 3.3 are in the form of powder, which can be compacted to various densities depending on the compaction pressure employed, but as can be seen in Table 3.4 increasing the density considerably raises the detonation velocity. Other explosives such as PETN, RDX, pentolite and tetryl are cast or pressed into various shapes, but frequently they are produced in the form of pellets or short cylinders. Plastic explosives, such as C-3 which consists of 77% RDX and 23% explosive plasticiser, can be moulded by hand to any desired shape. C-3 has a detonation velocity of 7.7 km/s. Yet another form of explosive which is used in some explosive welding operations is sheet explosive, which is mostly PETN explosive combined with a plasticiser. This is produced by the Nobel Explosives Co. Ltd. under the trade name

TABLE 3.3

Composition of Some Typical Explosive Mixtures

Explosive	Ingredients % by weight											Physical Form
	Explosive			Oxidiser		Fuel			Miscellaneous			
	NG	TNT	PETN	Amm. Nit.	Sod. Nit.	Carbon-aceous	Al.	Fuel Oil	Sod. Chl.	Amm. Chl.	Other	
Anfo				94				6				Granular
Trimonite 1		10		80			10					Powder
Trimonite 3		20		80								Powder
Ammonal		12		85			5					Powder
NG powder (Polar Rockite)	10			80		10						Powder
Permissable dynamite	9			67	5	9			10			Powder
Polar viking	8.4			70.7		8.8			10		2.1	Granular
Carrifax	9			10	46.5	6				28	0.5	Powder
Pentolite		90/50	10/50									Cast or pressed

TABLE 3.4

Relative Power, Sensitiveness, and Detonation Velocity of Some Explosives (53, 54, 55)

Explosive	Relative Power % Blasting Gelatine		Torpedo Friction Test (1 kg at 80°) cm	Fall Hammer Test (0.5 kg) cm	Density g/cc	Detonation Velocity km/s
	Ballistic mortar	Trauzl test				
Blasting Gelatine 92% NG, 8% NC	100				1.55	7.9
TNT	67	67	80 - 120	> 200	1.64 1.00	6.95 5.01
PETN	97	114	35 - 40	60 - 80	1.67 1.00	7.98 5.55
RDX	100	114	10 - 20	25 - 30	1.80 1.00	8.75 6.08
Tetryl	84	86			1.70 1.00	7.56 5.60
50/50 Pentolite	85	82			1.67 1.00	7.47 5.48
Gelignite			40 - 60	5 - 10		
TNT powder			100 - 200	160 - 200	1.1	4.85
NG powder			> 150	20 - 30	1.0	4.25
Metabel	80					7.00

of Metabel in sheets 250 x 250 x 3.2 mm, and by the E.I. Du Pont and de Nemours and Co. Inc. under the trade name of Detasheet which can be obtained in various thicknesses. Metabel and Detasheet have a detonation velocity of about 7 km/s. Lastly, mention must be made of detonating fuse, which has various applications in explosive welding processes, in addition to being used in the Dautriche method of measuring detonating velocity. Detonating fuse normally consists of a core of PETN explosive enclosed in a strip of paper formed into a tube, which is wrapped with a textile yarn and then covered with plastic. The Nobel Explosive Co. Ltd. manufacture this under the trade name Cordtex detonating fuse, which has a diameter of 5 mm and contains 10 g/m of PETN, while E.I. Du Pont de Nemours and Co. Inc. market a range of detonating fuse. Cordtex detonating fuse has a detonation velocity of 6.7 km/s.

A detonator is required to initiate detonation in secondary explosives, and for some less sensitive secondary explosives a booster charge is required in addition to the base charge in the detonator. Normally the detonator consists of a thin-walled copper or aluminium tube which is closed at one end, and partially filled with a charge of primary high explosive with a base charge of some secondary explosive. The primary charge can be lead azide or according to Taylor and Gay (59) it consisted, at that time, of a 2:1 mixture of lead azide and lead styphnate with 3% powdered aluminium, which is known as ASA. With a plain detonator, shown in Fig. 3.8a, the end of the safety fuse is inserted in the end of the detonator tube, and secured by crimping; the 'spit' of the safety fuse is adequate to initiate detonation in the primary high explosive. An alternative, which is normally used in explosive welding, is the electric detonator shown in Fig. 3.8b. In this case the fuse head consists of a bridge wire embedded in a flashing composition, which readily ignites when an electric current heats the wire, and this in turn causes the primary charge to detonate. The electrical energy is provided by an exploder, which is either a dynamo or magneto operated source, or now more commonly by a unit which is battery operated and includes a circuit tester facility to check the electric circuit prior to detonation.

The base charge in detonators is frequently of PETN, but

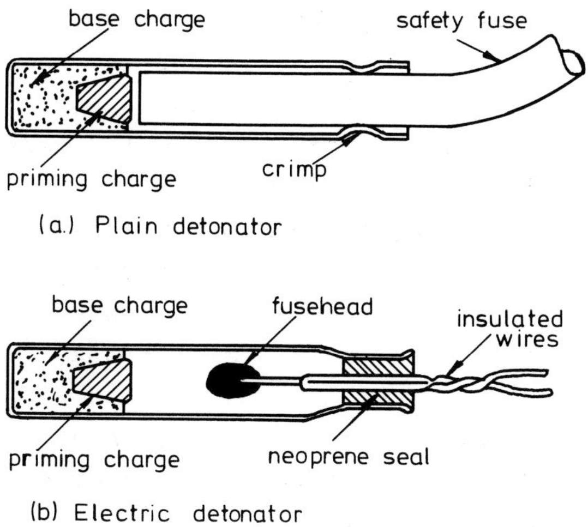

FIG. 3.8 Detonators

for higher temperature applications RDX is used. Originally the power of detonators was increased by a two stage compaction of the secondary charge; the first part of the base charge was compacted under a much higher pressure than the second part, which was compacted under the normal pressure. Such detonators were known as Briska-type detonators, but the same effect has been achieved in Britain by simply increasing the base charge; these detonators are known as star detonators. Only four detonators are now in common use, and Table 3.5 gives the mass of the base charge in each of these detonators.

Another variation between detonators is in the characteristics of the fuse head, for example, the Nobel Explosives Co. Ltd. produce a range of fuseheads, the most sensitive of which has a resistance of 1.0 - 1.8 ohms and the least sensitive 0.03 - 0.05 ohms. The least sensitive requires a firing impulse which is a thousand times greater than for the most sensitive fusehead, and it consequently gives much greater security against near lightning strikes, stray currents, galvanic currents, static electricity and electromagnetic radiation. Though detonators are extremely reliable, they can also be provided to a more stringent aerospace specification. These detonators are not only more carefully controlled and inspected during manufacture, but they also have a glass to metal seal instead of a neoprene seal, and they are also radiographed on completion as a final check.

TABLE 3.5

Charge Mass of Detonators

Type	Mass of Base Charge g
No. 6	0.24
No. 6 star	0.40
No. 8	0.45
No. 8 star	0.80

In explosive welding the detonation velocity of the explosive used is of considerable importance. The detonation velocities quoted in Table 3.4 are the maximum values, but in many explosives the detonation velocity is a significant function of diameter in cylindrical charges, or thickness of the layer of explosive as used in explosive cladding. This variation of detonation velocity is particularly of importance with the powder explosives used in explosive cladding operations. The detonation velocity may also be influenced by the particle size and by the addition of inert material to the explosive. Some data is available to indicate the influence of these variables on the detonation velocity of some explosives.

Much of the data on the effect of size on detonation velocity are on cylindrical charges, which are used in bore holes in mining and quarrying operations. However, in a bore hole the charge is much more confined than when it is in the form of a layer resting on a flyer plate, as in explosive cladding. Consequently, the data on cylindrical charges cannot be directly related to the detonation of uniform layers of explosive. This is demonstrated by the work of Shribman and Crossland (60) and Crossland and Williams (61), who measured detonation velocity in layers and in hollow cylindrical charges of Trimonite 1 explosive, the composition of which is given in Table 3.3. In Fig. 3.9 the detonation velocity is plotted as a function of radial thickness or the thickness of the layer, and it will be seen on this basis of comparison that the detonation velocity

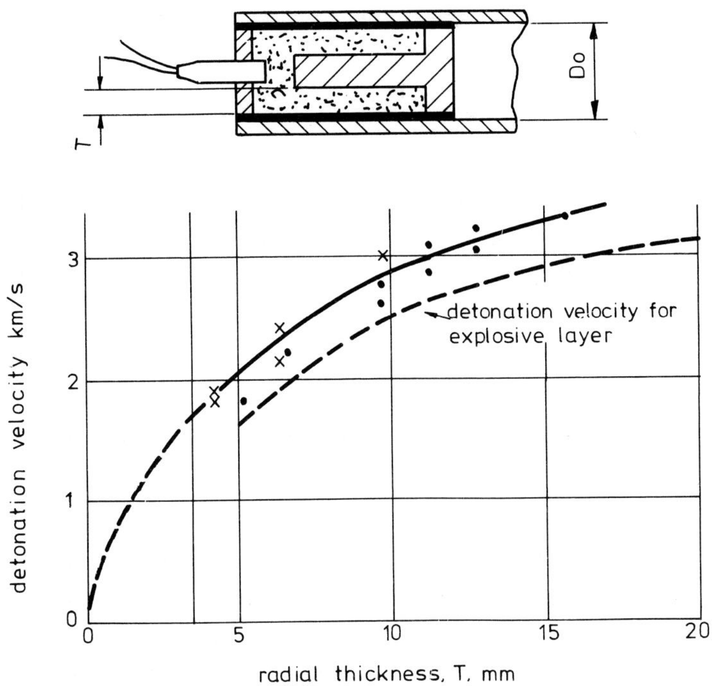

FIG. 3.9 Detonation in hollow cylindrical charges of Trimonite 1 (61) • D_o = 32 mm, x D_o = 19 mm

in hollow cylindrical charges is greater than in corresponding layer charges. In Fig. 3.10 data are shown for the detonation velocity in layer charges of Trimonite 1 in both a powder and a granular form and for Trimonite 3 powder. The difference between the detonation velocity of Trimonite 1 in the powder and granular forms is wholly attributable to the difference in density. Shribman and Crossland (60) also investigated the detonation velocity of Metabel sheet explosive in the thickness range from 3 to 13 mm, and they found no significant variation in detonation velocity over this range.

Anfo is an explosive with a low detonation velocity, which has been successfully used for large plate cladding. It is an explosive which, since 1956, has made considerable inroads into the sales of explosives for mining and quarrying. In 1972 Anfo accounted for 60% of the total explosives used in the USA. Extensive information on Anfo is available in handbooks (62), (63), but this relates to cylindrical charges in bore holes. Crossland and Cave (64) have provided some data on the detonation of layer charges of Anfo.

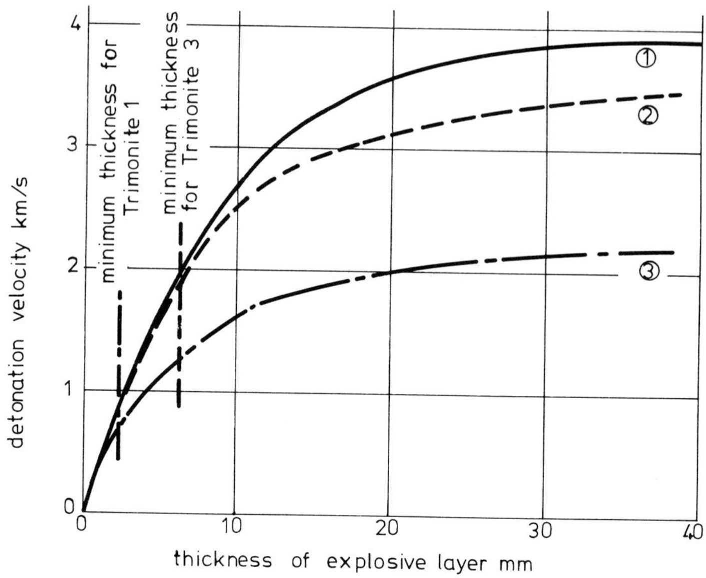

FIG. 3.10 Effect of thickness of explosive layer on detonation velocity (60). (1) Trimonite 3 powder ρ = 0.98 g/cc (2) Trimonite 1 powder ρ = 1.1 g/cc (3) Trimonite 1 granules ρ = 0.712 g/cm

Anfo is a mixture of porous ammonium nitrate prills and fuel oil. Theoretically the amount of fuel oil for oxygen balance is 5.7% by weight, and in practice 6% is commonly used, though Crossland and Cave (64) examined mixtures up to 20%. However, 15% of fuel oil appeared to saturate the prills and above 15% detonation was increasingly erratic. Anfo may be mixed with a concrete mixer for large charges, or manually by shaking the mixture contained within a plastic bag. Up to a fuel oil content of 15% the mixture remains as a free running coarse powder, which is easy to handle. If the explosive after mixture is not kept in sealed containers, there will be a steady evaporation of the fuel oil, so it is necessary to use the mixture within 24 hours if a reasonable control of the fuel oil content is desired. To achieve stable detonation within a short distance of the point of initiation of detonation, it is necessary to have a booster charge of some high detonation velocity explosive, for instance, a PETN pellet or some plastic explosive. Fig. 3.11 shows the variation of detonation velocity with thickness of the layer charge of 400

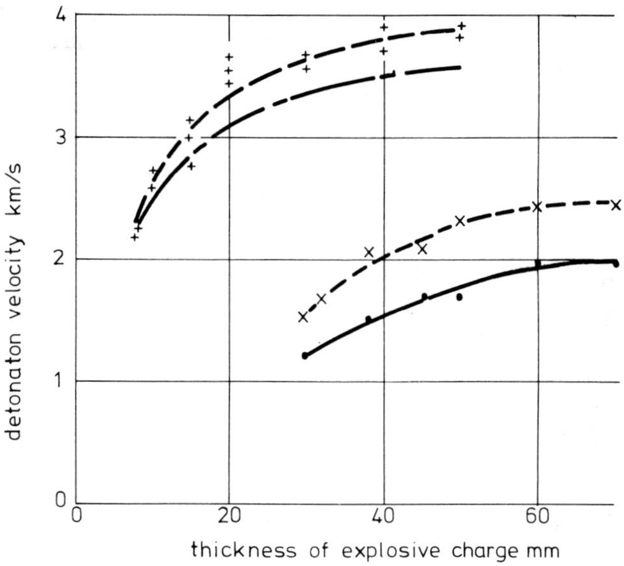

FIG. 3.11 Effect of charge thickness on detonation velocity of Anfo and Trimonite 1 (64). -•- Anfo with 12% fuel oil, --x-- Anfo with 6% fuel oil, -+- Trimonite 1, -·- Trimonite 1 (60)

mm x 150 mm. It will be noted that the 12% fuel oil mixture gives a consistently lower detonation velocity than the 6% mixture, and the minimum layer thickness for reliable detonation is 30 mm. At the same time as these tests were carried out, tests on Trimonite 1 were also carried out to repeat the tests reported by Shribman and Crossland, which had been carried out on smaller charges. These results are also shown plotted in Fig. 3.11. Tests to examine the effect of the percentage of fuel oil were carried out for a fixed charge thickness of 38 mm, and the results are shown in Fig. 3.12.

The data on the detonation velocity of Anfo layer charges are significantly different to those for confined cylindrical charges; the detonation velocity is much lower, which may be caused by the lack of confinement, and the minimum thickness for stable detonation in layer charges appears to be appreciably less. However, the results of Crossland and Cave (64) appear to be reproducable, and based on these data several large and medium size cladded plates of various combinations have been produced.

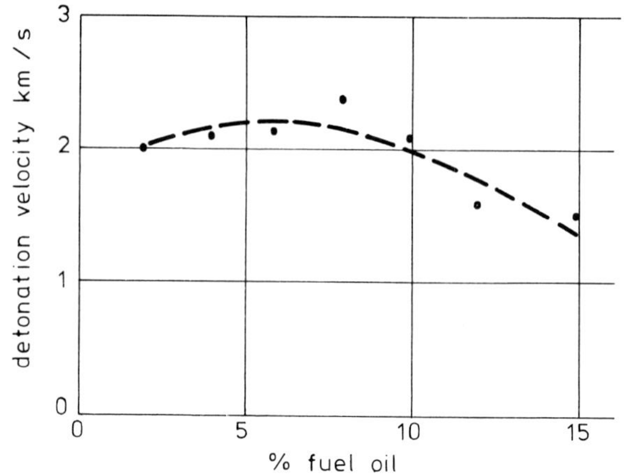

FIG. 3.12 Variation of detonation velocity of Anfo with percentage of fuel oil for a 38 mm thick layer (64)

The detonation velocity of explosives can be varied by varying the density and particle size. In the case of Anfo the detonation velocity might be reduced by increasing the porosity of the ammonium nitrate prills, and conversely it could be raised by reducing the porosity and reducing the particle size. According to Taylor (51) the most effective method of reducing the detonation velocity in explosives is to load the explosive with a non-reactive material such as common salt. Fig. 3.13 shows the variation of detonation velocity of

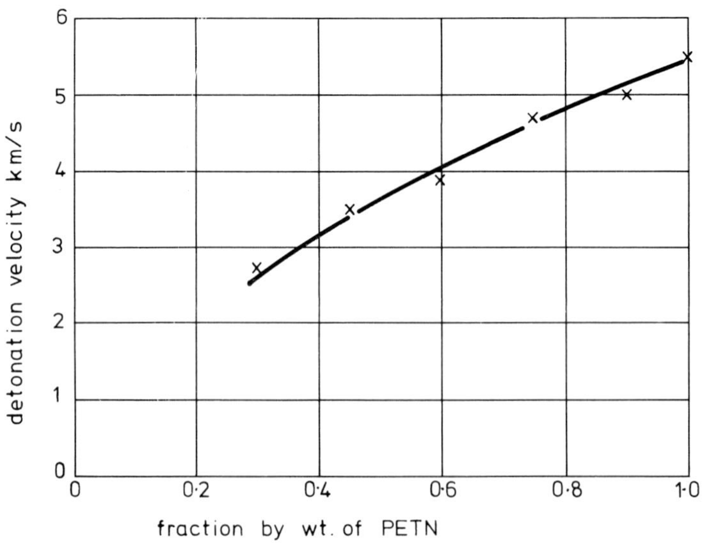

FIG. 3.13 Variation of detonation velocity of PETN/NaCl mixtures with fraction of PETN for a density of 1g/cc (51)

PETN/NaCl mixtures with the fraction of PETN and Fig. 3.14 shows similar data for NG/NaCl mixtures. No information appears to be available on the effect on the minimum size of charge which can be detonated as the fraction of non-reactive material is varied. There is also no published information on the effect of adding non-reactive materials to the lower detonation velocity explosive mixtures, such as TNT and NG powder explosives or Trimonite 1 or Anfo, though it is an established method for lowering the detonation velocity in practice.

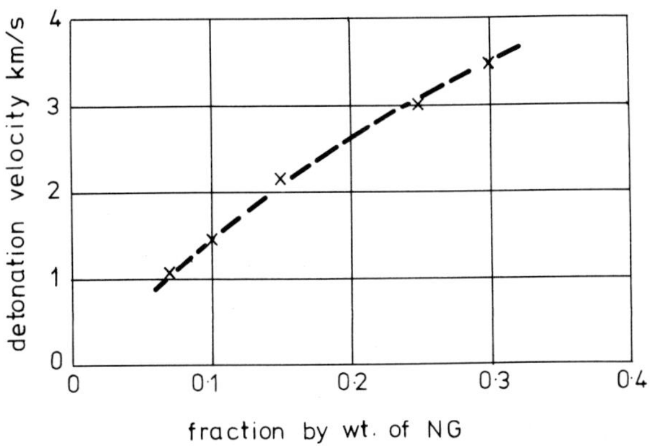

FIG. 3.14 Variation of detonation velocity of NG/NaCl mixtures with fraction of NG for · density of 1.35 g/cc (51)

The use of explosives in industry has legal problems, and the very considerable problems of safety. In the United Kingdom the manufacture, storage and transporting of explosives are governed by the Explosives Act of 1875 (65) as amended by the Explosives Act of 1923 (66), and relevant Orders in Council and Statutory Instruments and Regulations made thereunder. Furthermore, H.M. Inspector of Explosives have wide powers for making orders to deal with individual situations as they arise. Besides these specific laws and regulations relating to explosives, there is the requirement to satisfy the Health and Safety at Work Act, and various environmental and planning requirements. There are corresponding laws and regulations which have to be satisfied in other countries. Obviously, any explosive welding operation must be set up around these laws, orders and regulations.

The safety aspect of handling explosives is obviously most important. Though the energy required to initiate explosives and detonators is sufficiently great to give an adequate margin of safety, nevertheless it must be recognised that the necessary energy for initiating explosives or detonators may conceivably occur in several ways, such as impact, friction, heat or electrically. Adequate precautions against these possibilities must be taken; a single mistake could be fatal. It is essential to produce a code of practice for the safe handling of explosives, and to ensure that all personnel working with explosives are thoroughly familiar with the code. Furthermore, personnel involved in the operation should have had an adequate training in the general use and handling of explosives, and in the particular problems associated with explosive welding. Refresher courses should be organised in order to maintain a sense of awareness of the dangers associated with explosives, as familiarity tends to lead to carelessness in observing the rules laid down. A high degree of discipline is required on the site with a very well defined line of command. More details of safety instructions and requirements are given by Fordham (54), Veil (67), Thomas (68) and Ezra (69).

The noise pollution problem particularly associated with large plate cladding, poses serious limitations on the location of sites. This has forced some operators to carry out their cladding work on sites in remote areas, with all the consequential cost of transportation of the materials to be cladded and the explosive to the site, and the removal of the product from the site. It will also probably involve the provision of an access road and services to the site. Other operators have overcome this, to some extent, by carrying out the work in underground chambers, but this would be very expensive unless suitable underground workings are readily available; for example, in their Chikushino plant the Asahi Chemical Co. Ltd. have made use of a large underground complex constructed during the Second World War. However, it is necessary to ensure that such chambers are sufficiently deep and located in a suitable geological structure, in order to prevent a blowout or a collapse of the roof of the chamber; there is at least one example of the failure of such a chamber, which was too close to the surface. Yet another possibility is to construct a reinforced steel chamber which is then embedded in a suitably designed reinforced concrete structure and covered by a thick layer of earth as shown in Fig. 3.15. Ventilation ducts with fans are provided. Yamamoto (70) describes such a chamber with a inner diameter of 9600 mm and length 10000 mm, which he claims reduces the noise of 100-140 kg of explosive at 500 metres from 118-130 db without a chamber to 84-96 db. A chamber of this general design has been constructed at the Aibano Works of the Asahi Chemical Co. Ltd. and it is used for the commercial production of clad plate.

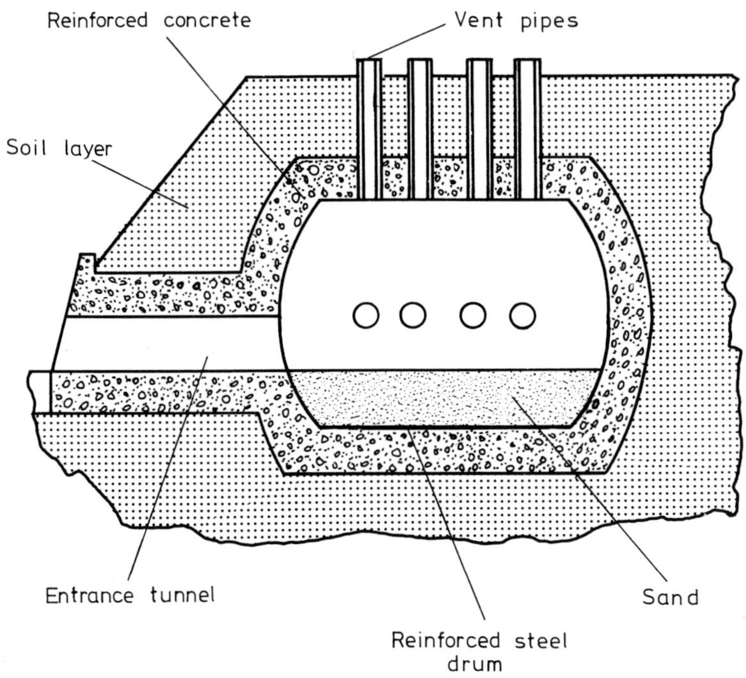

FIG. 3.15 Explosive welding chamber (70)

An alternative solution to the problem of noise and the damaging effect of the shock wave is to bury the large plate cladding set-up under soil or to submerge it under water, though ingress of soil or water into the gap between the flyer and parent plates must be avoided. It is known that the damaging effect of terrorist bombs can be significantly reduced by covering the bomb with a plastic envelop filled with water. However, if such a system could be adopted for explosive welding it would also have the effect of providing some confinement to the explosive, with a resulting increase in the detonation velocity and the velocity imparted to the flyer plate for a given thickness of explosive.

The problem of noise with much smaller charges associated with tube to tubeplate welding or plugging of heat exchangers is much less serious, and with suitable precautions it is possible to carry out such processes within a factory environment.

4

INTERACTION BETWEEN EXPLOSIVES AND THE WORKPIECE

In explosive welding the pressure produced in the region of the detonation front in the explosive charge is used to provide rapid acceleration of the flyer plate to a high velocity, prior to it impacting the parent plate. From Chapter 2 it will be realised that one of the important parameters in explosive welding is the magnitude of the velocity imparted to the flyer plate by the explosive charge. Additionally it is of interest to know the distance required by the flyer plate to accelerate to its terminal velocity, in order to ensure that the initial stand-off is adequate as otherwise an excessive charge would be necessary to produce the required velocity.

The pressure produced in the detonation front will be transmitted into the flyer plate as a stress wave. If the pressure at the peak of the stress wave is sufficiently large then the associated wave velocity will be greater than at a lower pressure because of the reduction in compressibility. As illustrated in Fig. 4.1 this difference in velocity leads

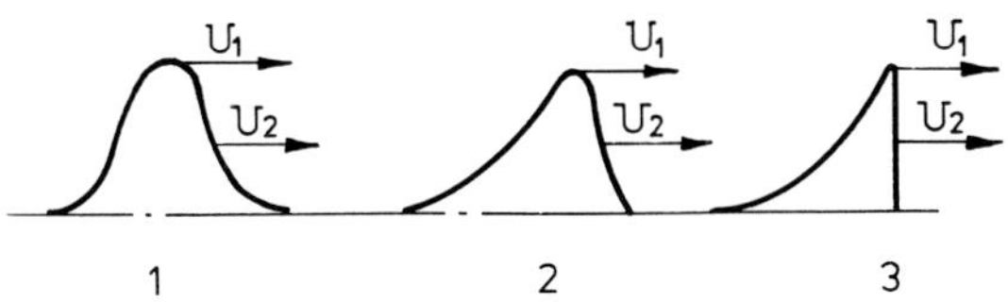

FIG. 4.1 Development of a shock wave $V_1 > V_2$

to the generation of a sharp fronted stress wave or what is known as a shock wave. Normal elastic wave theory is not applicable to such sharp fronted waves, in which there are large pressure and temperature gradients and a considerable dissipation of energy. As a shock wave travels through the material its energy will be dissipated and it will ultimately reduce to an elastic wave and become less sharp fronted. A shock wave will be reflected from a free surface as a tensile wave, but using the arguments adduced for the generation of a shock wave it can be readily appreciated that tensile shock waves cannot be generated. Consequently the reflected tensile wave will broaden out into a normal stress wave. Tensile waves generated from reflection of a shock wave, especially in regions of discontinuity such as a corner, can lead to a tensile failure or what is known as *spalling*. The portion of the plate which becomes detached, or what is known as the *scab*, flies off at a high velocity.

With the low detonation velocity explosives used in many explosive welding applications the intensity of the pressure generated in the flyer plate is such that either a shock is not produced, or if it is produced it is a weak shock wave which quickly reduces to an elastic wave. However, more intense pressures are generated in the region of impact between the flyer and parent plates, and shock waves may be initiated from this region.

There are many authoritative books written on the subject of stress transients in solids such as Kolsky (71), Rinehart and Pearson (53) (72), Goldsmith (73), Shewman and Zackay (74) Johnson (75), Wasley (76) and Rinehart (77). However, the subject is of such importance in explosive welding, and there also appears to be some confusion in the literature on explosive welding relating to the velocity of wave propogation, that it is necessary to give a brief resumé.

Low intensity stress waves will propagate through solids at the velocity of sound within the solids. The velocity of sound can be readily deduced from elastic theory, for instance see Johnson (75), given the boundary conditions to be satisfied. There are essentially four cases to be considered. For a thin rod in which the effect of transverse strain and inertia are neglected the velocity of the longitudinal wave is given by:

$$C_R = \left(\frac{E}{\rho}\right)^{\frac{1}{2}} \qquad (4.1)$$

where E is the Young's modulus and ρ the density. In a rectangular bar or plate, which has a width and length great compared with the thickness, the velocity of a plane wave travelling down the bar will be:

$$C_P = \left\{\frac{E}{\rho(1 - \nu)^2}\right\}^{\frac{1}{2}} \qquad (4.2)$$

where ν is Poisson's ratio. For a uniform bar which is constrained to have zero transverse deformation the velocity of longitudinal wave propagation, which is sometimes known as a dilalational wave, will be

$$C_L = \left\{\frac{E(1 - \nu)}{\rho(1 + \nu)(1 - 2\nu)}\right\}^{\frac{1}{2}} \tag{4.3}$$

Lastly if a thin cylindrical tube is subjected to a torsional disturbance then the velocity of transmission of a torsional or shear or distortional or transverse wave will be:

$$C_T = \left(\frac{G}{\rho}\right)^{\frac{1}{2}} \text{ or } \left\{\frac{E}{2\rho(1 + \nu)}\right\}^{\frac{1}{2}} \tag{4.4}$$

where G is the shear modulus.

Some of the published values of these wave velocities seem to have been calculated using values of the elastic constants, E, G and ρ, which are not mutually compatible. Table 4.1 gives values of these wave velocities based on values of elastic constants given by Kaye and Laby (78).

Table 4.1

Velocity of Elastic Waves in Metals km/s

Material	C_R	C_P	C_L	C_T
Aluminium	5.105	5.438	6.398	3.110
Brass 70/30	3.435	3.646	4.351	2.091
Copper	3.812	4.058	4.758	2.325
Lead	1.191	1.327	2.145	0.702
Magnesium	5.071	5.301	5.813	3.155
Nickel (hard)	4.961	5.211	5.686	3.069
Steel (mild)	5.196	5.431	5.955	3.236
Tantalum	3.338	3.552	4.157	3.338
Titanium	5.066	5.349	6.070	3.117
Zinc	3.898	4.025	4.266	2.466

These equations for elastic wave velocities and the values

given in Table 4.1 are based on the assumption that the velocity of wave propagation are solely dependant on the elastic constants, which are assumed to be independant of the magnitude of the compressive pulse. However, when the pressure is very high as in an intense shock wave, then the shear strength becomes negligible compared with the pressure and the material behaves as a fluid. In a fluid only one type of wave is possible, that is a compressive wave which will have a velocity:

$$C_F = \left(\frac{dP}{d\rho}\right)^{\frac{1}{2}} \qquad (4.5)$$

which at low pressure will reduce to:

$$C_F = \left(\frac{K}{\rho}\right)^{\frac{1}{2}} \qquad (4.6)$$

where K is the bulk modulus which is given by:

$$K = \frac{E}{3(1 - 2\nu)} \qquad (4.7)$$

However, under high pressure the shock velocity will be given by Equation 4.5 and calculation and experiment, for instance see Walsh et al (79), indicates that the shock velocity, U, is linearly related to the particle velocity, u, or,

$$U = a + bu \qquad (4.8)$$

where a and b are constants. These constants have been established experimentally. However, it might be expected that the constant a should be equal to the compressive wave velocity given by Equation 4.6, as at small particle velocities the pressure is correspondingly small. Table 2 gives some values of the constants a and b provided by Evans (80) with corresponding data for C_F calculated from elastic constants given by Kaye and Laby (78). The correspondence between a and C_F is reasonable bearing in mind that a is obtained from the best fit line through data points at very high pressure. With tantalum the discrepancy may well be due to the fact that $1 - 2\nu$ in Equation 4.7 is very small and subject to error.

Table 4.2

Material	a km/s	b	C_F km/s
Aluminium	5.26	0.70	5.293
Brass 70/30	3.47	1.69	3.620
Copper	3.99	1.50	3.927
Lead	{2.03 2.30	{1.52 1.27	1.986
Magnesium	4.49	1.27	4.529
Nickel (hard)	4.65	1.45	4.598
Tantalum	3.13	1.56	3.428
Titanium	4.78	1.09	4.888
Zinc	3.71	1.45	3.176

Fig. 4.2 illustrates the case of an explosive charge in

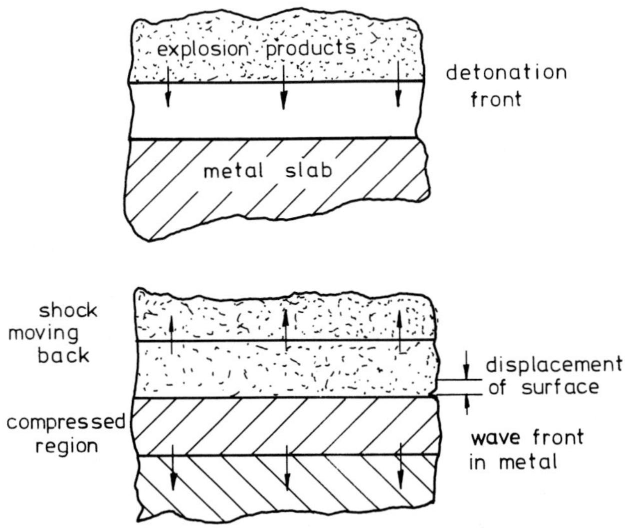

FIG. 4.2 Interaction at metal/explosive interface

intimate contact with a metal slab, in which the detonation front is assumed parallel to and propagating towards the metal surface. It is assumed that both metal and explosive are semi-infinite in extent. When the detonation front reaches the metal surface a sharp fronted compressive stress wave is induced in the metal. Ahead of the stress wave the metal is undisturbed, but behind the wave front the pressure is extremely high. The material at the wave front is highly compressed, and it is accelerated to a high particle velocity in the direction of the wave front. The pressure, density, temperature, particle velocity and stress wave velocity are governed by equations identical with Equations 3.1, 3.2 and 3.3, except that E_o does not include a chemical release of energy component, together with the equation of state for the particular metal. From Equations 3.4, 3.5 and 3.6 it can be shown that the pressure behind the wave front is given by:

$$P \approx p_1 - p_o = \rho_o V_s u_1 \qquad (4.9)$$

where V_s is the velocity of the wave front, u_1 is the particle velocity behind the wave front and ρ_o is the density in front of the wave front. The implies that the particle velocity varies with time in a similar way to the variation of pressure with time.

At the explosive/metal interface the pressure and particle velocity will be the same each side of the interface, at the instant that the detonation front arrives at the interface. Duvall and Fowles (81) have used this consideration to determine the conditions at the explosive/metal interface, for a number of explosive/metal combinations, by plotting curves of pressure versus particle velocity for various metal and explosives. The curves are shown in Fig. 4.3 and the points

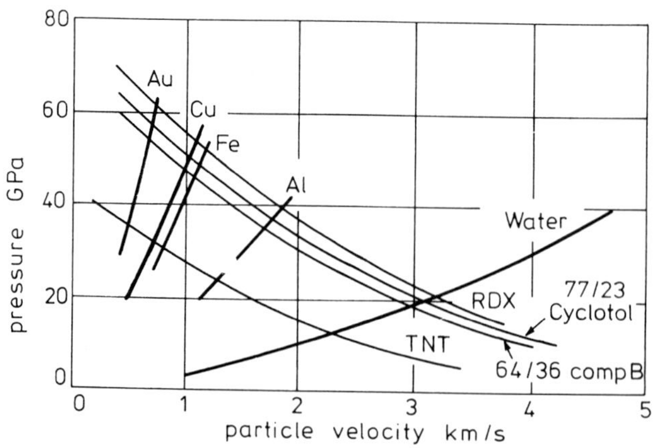

FIG. 4.3 Pressure/particle velocity data for several metals and explosives (81)

of intersection indicate the conditions when the explosive represented by one of the curves is detonated in intimate contact with the metal represented by the other curve. For example, for aluminium in contact with TNT the pressure is 24 GPa and the particle velocity is 1.25 km/s.

It can be appreciated from the example of aluminium in contact with TNT that a very high pressure is generated in the metal. However, as the stress wave travels through the material there will be a decay of the peak pressure and in order that momentum is conserved the pulse length must increase. Jones (82) discusses this problem in greater detail and Fig. 4.4, which illustrates this attenuation of the peak pressure, is taken from that paper. In order to prevent damage of the metal surface by the very high peak pressure a buffer, for example, of rubber or plastic, may be interposed between the explosive and the metal surface, which will cause more rapid attenuation of the peak pressure. Buffers are used in explosive welding particularly for high detonation velocity explosives, such as Metabel and Detasheet sheet explosives. In explosive welding it is usual, however, to use a low detonation velocity explosive which creates a much lower pressure at the metal/explosive interface than a high detonation velocity explosive.

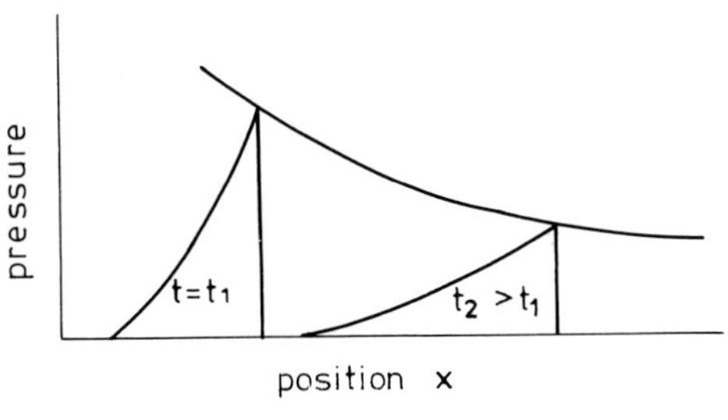

FIG. 4.4 Attenuation of the peak pressure

In order to examine the motion imparted to a metal plate of finite thickness the action of a rectangular elastic compression wave will be considered, and it will be assumed that its length is less than the thickness of the plate. Fig. 4.5a shows the form of stress wave and also the particle velocity distribution. As the wave passes into the front surface of the plate, the surface is displaced by a distance equal to the

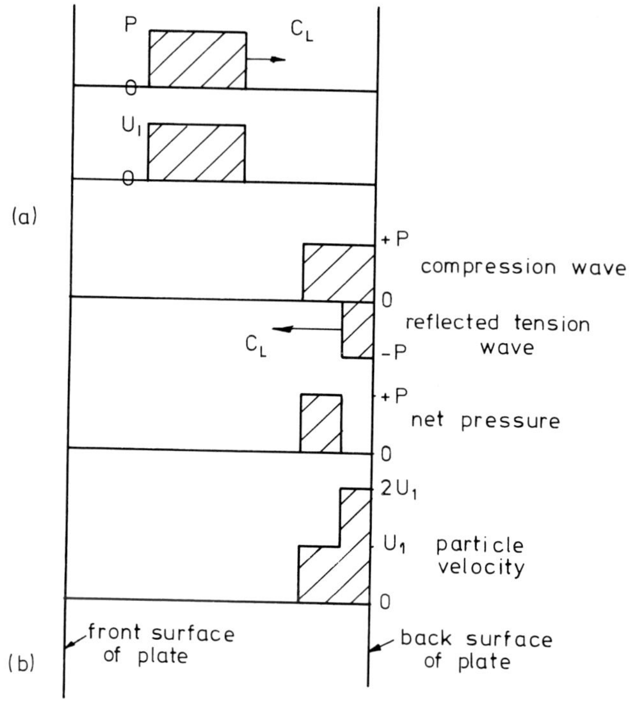

FIG. 4.5 Reflection of compressive stress wave from back surface of the metal plate

time integral of the particle velocity over the duration of the wave, that is $_o\int^t u_1 \, d\, t$, which in this example would be u_1t, where t is the duration of the wave. When this compression wave reaches the back surface of the metal slab, it is reflected back as a tension wave as shown in Fig. 4.5b. However, due to the change in direction of the wave and its change of sign, it can be seen from Equation 4.9 that the particle velocity of the reflected tension wave is still acting to the right. Consequently, for the duration of the wave the back surface of the flyer plate experiences a particle velocity of $2u_1$, and as a result it will be displaced a distance $2u_1t$. The tension wave reaches the front surface, where it is reflected as a compression wave, and the front surface is displaced a further increment $2u_1t$. In the absence of any internal or external dissipative forces, this bouncing to and fro between the front and back surfaces will continue ad-infinitum, and each time the wave reaches the back surface the surface will be displaced $2u_1t$. The variation of velocity and displacement of the back surface are shown in Fig. 4.6.

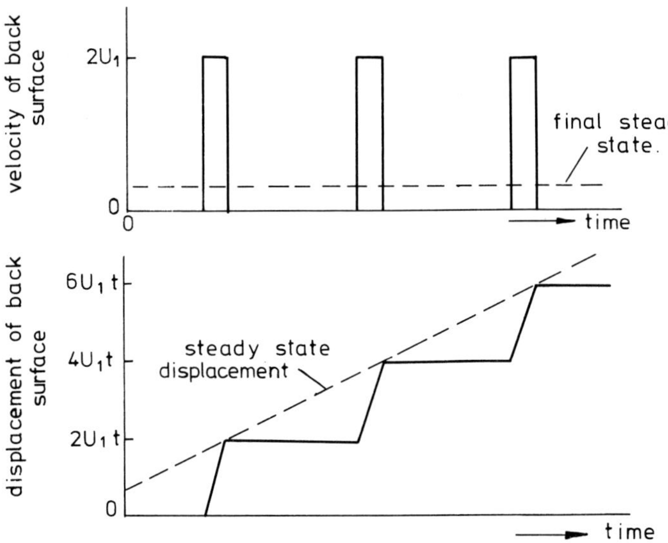

FIG. 4.6 Velocity and displacement of back surface

The ideal situation described in Fig. 4.5 and 4.6 does not hold in practice, not only are there internal and external dissipative forces, but the rectangular wave form assumed is unrealistic. For a given explosive/metal combination the peak pressure will be the same irrespective of the thickness of the explosive charge, as indicated in Fig. 4.7. Also the length

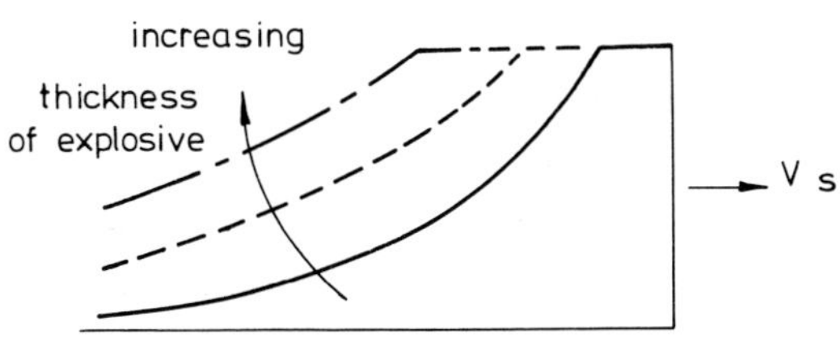

FIG. 4.7 Compressive stress wave induced by explosive loading

of the pressure pulse is, in general, longer than that shown in Fig. 4.5a. For convenience it will now be assumed that the stress wave is triangular and sharp fronted as indicated in Fig. 4.8. As the compression wave is progressively reflected as a tension wave there is an increasing net tensile stress as

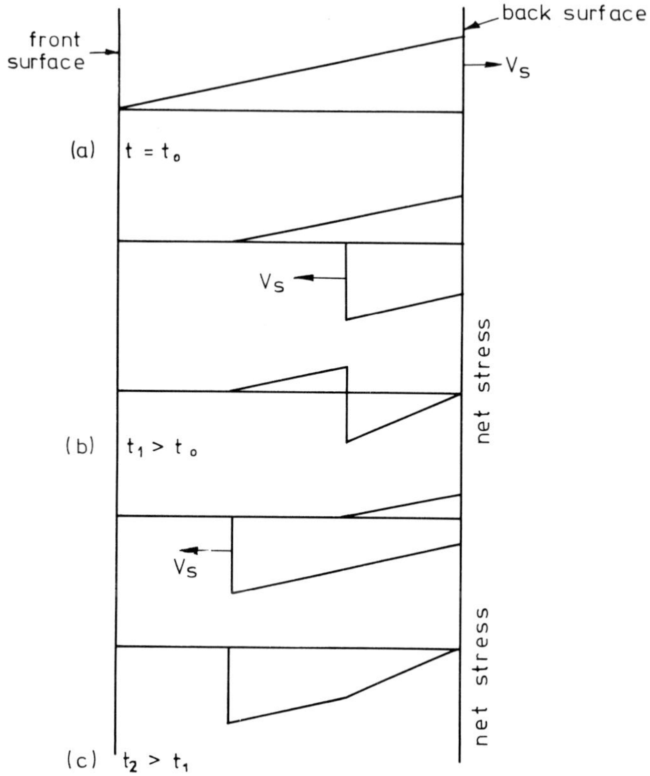

FIG. 4.8 Reflection of a triangular compressive stress wave in a plate

the front of the tension wave gets further from the back surface, until it reaches a maximum value numerically equal to the peak compressive stress, assuming that there are no internal dissipative forces. If, at a particular distance from the back surface the net tensile stress exceeds the fracture strength of the material, then failure will occur and a scab will be released. This scab will contain the kinetic energy associated with the particle velocities contained within it.

If there are no external dissipative forces the system will be conservative, and the momentum imparted by the explosive pulse will be conserved. However, due to internal dissipative forces the peak of the stress wave will decrease, while its wave length will increase to satisfy the conservation of momentum. Ultimately all the differences in particle velocity will be damped out, and the whole plate will possess a uniform particle velocity or will have achieved a steady state velocity. If the compressive wave is short compared with the thickness of the plate, then the peak velocity imparted to the back surface, when the first compressive stress wave is reflected from it as a tension wave, will be greater than the steady state velocity. If the velocity of the back surface was to be recorded, it would be seen to consist of a succession of velocity pulses which decrease in amplitude and increase in wave length with time until a steady state velocity is reached. With the explosive charges required in explosive welding the wave length of the stress wave is large compared with the plate thickness. Consequently, when the first reflected tension wave arrives at the front surface and it is reflected back as a compressive wave, this will add to the compressive stress still being applied to the front by the gaseous products behind the detonation front. The velocity of the back surface is protrayed in Fig. 4.9.

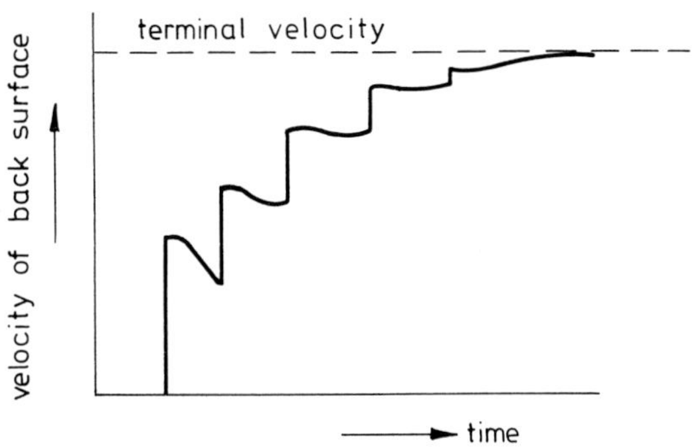

FIG. 4.9 Velocity imparted to a plate by an explosive charge

So far the discussion has been devoted to a plane detonation wave impacting the surface of a plate at normal incidence. In explosive welding, however, the detonation front runs parallel to the surface of the flyer plate. The duration of the pressure will be short, as the explosive products, being free to expand, do so rapidly. The thicker the layer of explosive the longer the pressure will be substained, even though the peak pressure is unaffected by the thickness of the charge. The pressure induced in the metal surface for sweeping incidence is appreciably less than for normal incidence. For instance, Duvall and Fowles (81) show for aluminium in contact with composition B explosive that the interface pressure is 36 GPa for normal incidence and 18.9 GPa for sweeping incidence, and for iron the corresponding figures are 47.5 GPa and 19.6 GPa. Another important difference, as can be seen in Fig. 4.10 is that the pressure wave induced in the flyer plate is oblique, and when this is reflected from the back surface of the plate it will give rise to both a dilational wave and a shear wave.

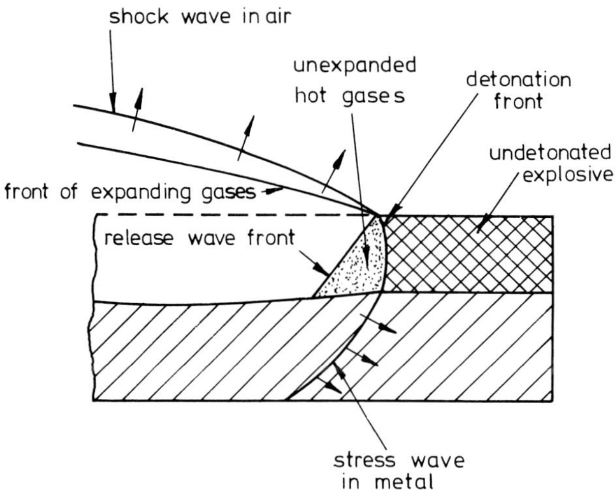

FIG. 4.10 Detonation of a layer of explosive in contact with a metal plate

From the foregoing it can be seen that the problem of calculating the flyer plate velocity is formidable, and various semi-empirical and empirical equations have been suggested, and some of these are reviewed by Shribman and Crossland (60). Perhaps the most extensively used equations are those based on ideas put forward by Gurney (83), and recently reviewed by Kennedy (84). Basically in the Gurney model it is assumed that the driven metal is projected in a direction normal to its surface, regardless of the direction of detonation propagation. Two further assumptions are made to obtain mathematical tractability and are crucial to the method. First, a specific energy, E, with a characteristic value for each explosive, is assumed to be converted from chemical energy in the initial state to kinetic energy in the final state. Secondly, this kinetic energy is partitioned between the metal and the gaseous products of detonation, by assuming a velocity distribution in the constant density products. For a symmetric configuration such as a cylinder or sphere or flat sandwich, that is an explosive layer sandwiched between two metal plates, the momentum balance requirement is automatically satisfied, so the energy balance equation fully defines the final state. For asymmetric configurations, such as open-faced sandwich or what is basically an explosively loaded flyer plate, an explicit momentum balance equation must be solved simultaneously with the energy balance equation.

It is worthwhile to consider the solution for the open-faced sandwich shown in Fig. 4.11. The velocity of the gas

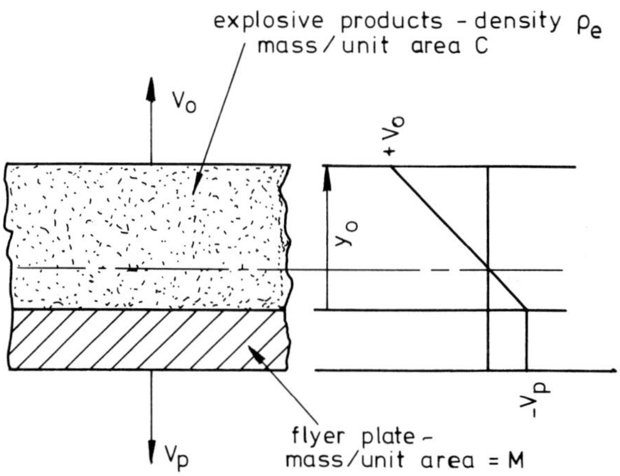

FIG. 4.11 Velocity distribution in open-faced sandwich

being expelled at the upper surface of the explosive is V_o, and at the interface between the gas and the flyer plate the velocity is Vp in the opposite direction. At some intervening position the gas velocity is zero. If the velocity profile is linear then the velocity as a function of y will be:

$$V = (V_o + V_P)\frac{y}{y_o} - V_P \tag{4.10}$$

Conservation of momentum leads to:

$$-M\,V_P + \rho_e \int_0^{y_o} \left[(V_o + V_P)\frac{y}{y_o} - V_P\right] dy = 0$$

which noting that $\rho_e y_o = C$, the mass of explosive per unit area, reduces to:

$$\frac{V_o}{V_p} = \frac{2}{R} + 1 \qquad (4.11)$$

where R is the mass of explosive for unit mass of flyer plate or C/M.

From energy considerations:

$$CE = \frac{1}{2} MV_P^2 + \frac{1}{2} \rho_e \int_0^{y_o} \left[(V_o + V_P)\frac{y}{y_o} - V_P \right]^2 dy$$

or

$$CE = \frac{1}{2} MV_P^2 + \frac{1}{6} C \frac{(V_o^3 + V_P^3)}{(V_o + V_P)}$$

which with Equation 4.11 leads to:

$$V_P = \sqrt{2E} \left\{ \frac{(1 + \frac{2}{R})^3 + 1}{6(1 + \frac{1}{R})} + \frac{1}{R} \right\}^{-\frac{1}{2}} \qquad (4.12)$$

Equation 4.12 appears to fit the experimental data reported by Shribman and Crossland (60) and Crossland and Cave (64) remarkably well. They used a pin contactor technique with an associated pulse circuit and a high speed recording oscillograph, to record the displacement of the flyer plate. The values of E for some of the typical explosives which can be used in explosive welding, and for which there is adequate if not comprehensive data, are given in Table 4.3, while data relating to other explosives are given by Kennedy (84).

The information in Table 4.3 is based on data given by Shribman and Crossland (60) and Crossland and Cave (64), who used a range of materials and thicknesses up to 10 mm for their flyer plates. Within this range of materials and thicknesses it appears that neither material nor thickness has any significant effect on the velocity imparted to the flyer plate. However, this must not be assumed to hold for much thicker plates, where the energy to plastically bend the flyer plate may form a more significant fraction of the available energy than would be the case for a thinner plate. It will be noted from Table 4.3, that there appears to be no simple relation between the calorific value of the explosive and E. Kennedy (84) concludes that for high density explosives which detonate ideally, the value of E is approximately 70% of the calorific value. For low density explosives which detonate non-ideally, the value of E is a much smaller percentage of the calorific value; for Trimonite 1 the percentage is 11% and for Anfo with 6% fuel oil it is only 5%.

Table 4.3

Values of E in Equation 4.12

Explosive	Density g/cm^3	E		Range of R on which E is based	Calorific Value kcal/kg
		kJ/kg	kcal/kg		
Metabel	1.47	3300	788	0 - 1.0	900-1050
Trimonite 1	1.10	582	139	0 - 2.4	1260
Trimonite 2	0.98	543	130	0 - 2.0	1034-1260
Anfo 6%	0.85	203	48.5	0 - 2.5	920
Anfo 12%	0.90	203	48.5	0 - 2.5	790

Fig. 4.12 shows a plot of the non-dimensional quantity $V_P/\sqrt{2E}$, from Equation 4.12, against the explosive loading R. The value of this non-dimensional quantity for a cylinder filled with explosive is given by Kennedy (84) as:

$$\frac{V_P}{\sqrt{2E}} = \left(\frac{1}{R} + \frac{1}{2}\right)^{-\frac{1}{2}} \tag{4.13}$$

and this is also plotted in Fig. 4.12. As might be hoped, Equation 4.13 fits the experimental data on tube velocity reported by Crossland and Williams (61) much better than Equation 4.12. However, the values of E for various explosives deduced from their data are appreciably different to those quoted in Table 4.3, as can be seen in Table 4.4. There is, of course, a significant difference between the two situations as

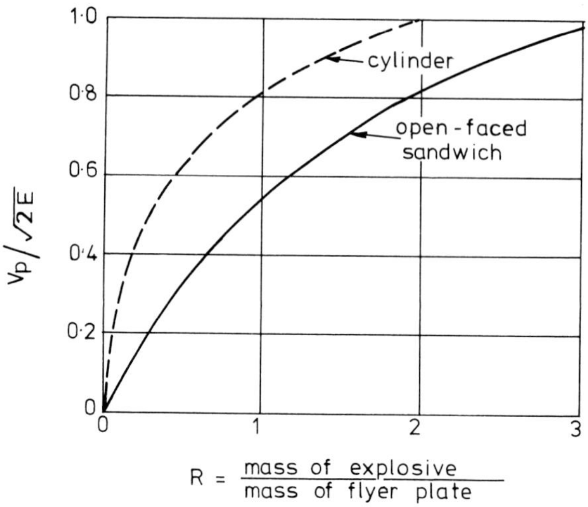

FIG. 4.12 $V_p/\sqrt{2E}$ for an open-faced sandwich and for an explosive filled cylinder

in the tubular configuration there is a continuous expenditure of energy in the plastic work in expanding the tube. This implies that there would be less energy available as kinetic energy of the tube, but calculations of the work required to expand the tubes used in the reported experiments suggest that it is negligible in comparison with the kinetic energy.

Table 4.4

Values of E in Equation 4.13

Explosive	Density g/cm^3	E kJ/kg	E (Table 4.3) kJ/kg	Calorific Value kcal/kg
Metabel	1.47	1342	3300	900-1050
Pentolite	1.6	1342	-	1280
Trimonite 1	1.1	277	582	1260

In some applications a radial inward velocity is imparted to a tube by an external cylindrical charge, as in implosive tube to tube welding or the compaction of metallic and ceramic powders contained in a metallic tube. In this situation it might appear that the velocity imparted to the tube would be more in accord with Equation 4.12 for the open-faced sandwich, but in this application the plastic work expended in contracting the tube increases rapidly with increasing diametral strain. Consequently, though the same peak radial velocity should be achieved as predicted by Equation 4.12, it will be followed by a more rapid deceleration.

Besides the need to know the terminal velocity of the flyer plate, it is important to know the flyer plate displacement needed to achieve this velocity. For a high detonation explosive such as Metabel sheet explosive, the pressure pulse will be of very large magnitude and of short duration. Consequently the flyer plate will be subject to a very high acceleration for a short time, and it will reach its maximum velocity within a short distance. With a low detonation velocity explosive, say with a detonation velocity less than 3-4 km/s, the magnitude of the pressure pulse will be much smaller, and to achieve the required impulse a much thicker layer of explosive will be required, which will prolong the duration of the pressure pulse. As a result the acceleration will be less but it will persist for longer and the distance for the flyer plate to reach its maximum velocity will be greater. An analytical procedure to describe the motion of the plate is made difficult by the problems of predicting the pressure/time relationship of the pressure pulse applied by the explosive to the flyer plate, and adequately describing the dissipative forces in the flyer plate which result in the decay of the stress waves acting in the plate. Additionally, there is the complexity of the stress waves induced in the flyer plate by the advancing detonation front, and their interactions.

There has been an unsubstantiated suggestion that three reflections of the initial compressive stress wave from the back surface of the flyer plate are required to achieve 90% of the terminal velocity. This would lead to a flyer plate displacement of less than half the flyer plate thickness. However, this suggestion appears to be no more than an inspired guess. Though the pin contactor measurements of Shribman and Crossland (60) does not provide a continuous measurement of displacement, nevertheless with several contactors arranged at different distances from the back surface of the flyer plate several points on the displacement/time curve can be determined. They concluded with Trimonite 1 that the terminal velocity was achieved in a distance less than the flyer plate thickness. Whereas Crossland and Cave (64) found for Anfo that the displacement to achieve the terminal velocity appeared to be less than twice the flyer plate thickness, but greater than that required for Trimonite 1, though more recent data suggest that a somewhat greater displacement is required.

Takizawa et al (85) have examined the problem of flyer plate acceleration by taking flash X-ray photographs of a flyer plate in flight. They examined the acceleration of 3 mm thick flyer plates of various metals using several explosives which had very different characteristics. It was noted for one of the explosives that there was a considerable distance of 200-300 mm from the point of detonation before stable conditions were achieved. This suggests that they had not employed an adequate booster charge to initiate stable detonation, which is essential for the lower detonation velocity explosives. For this same explosive it was found that the displacement to achieve the terminal velocity was up to 5 to 6 times the flyer plate thickness, which seems to be much greater than found by other workers. However, this may not have been a very representative explosive as it had a low density, which for a given explosive loading resulted in a very thick layer of explosive compared with the plate thickness. As a result the duration of the pressure pulse would be increased giving a longer period of acceleration.

Smith and Linse (86) used a slanted wire technique to continuously measure the displacement-time characteristics of a flyer plate for two types of granular ammonium nitrate based explosives. Their results for two aluminium flyer plate thicknessess of 6.3 and 19.1 mm and two explosive charge thicknesses of 12.7 and 25.4 mm show that the terminal velocity is achieved within a distance equal to the flyer plate thickness. The data reported compares well with that reported by Shribman and Crossland (60) and Crossland and Cave (64). From the limited data available it would seem for low detonation velocity explosives, such as Anfo, and for the explosive loading ratios used in explosive cladding, that the stand-off distance should be at least twice the flyer plate thickness to allow the terminal velocity to be achieved before impact. For higher detonation velocity explosives, such as Trimonite 1, it is probable that the stand-off might be reduced from twice the flyer plate thickness to the flyer plate thickness, and for even higher detonation velocity explosives, such as Metabel sheet explosive, the stand-off might be further reduced.

5

EXPLOSIVE WELDING PARAMETERS

It has been shown in Chapter 2 that under certain circumstances a high velocity jet may be formed at the contact point in a high velocity oblique impact between two metal surfaces, and it is commonly believed that such a jet is essential for welding to occur. The jet is imagined to contain the surface layer of oxides, adsorbed gases and other contaminants of both the flyer and parent plates. Consequently in the vicinity of the stagnation point virginly clean surfaces are brought together under very high pressure, and a solid state weld is produced in which there is equilibrium between the close range attractive and repulsive forces. In a cold pressure weld Shorshorov and Krasulin (87) consider that the process is one of obtaining intimate contact between two metal surfaces, followed by a stage of chemical interaction between the surfaces, which may be followed by a third stage when the strength is reduced by the formation of a brittle interlayer. Though there are considerable differences in the duration and contact pressure at the interface in a cold pressure weld and that in an explosive weld, nevertheless it is probable that the same basic mechanism holds in both cases. However, though this provides a simple physical explanation of the formation of a weld under high velocity oblique impact, it does not provide any guidance as to the essential parameters to be satisfied to provide an acceptable weld.

It is necessary firstly to define what is meant by an acceptable weld. One definition might be that the strength of the weld should be as strong as the weaker of the two component materials as defined by one or more of the mechanical tests discussed in Chapter 9. The problem with that definition is it requires a destructive test, which does not provide an overall assessment of the integrity of the weld. An alternative definition might be that a certain percentage of the area of the weld should be satisfactorially bonded as determined by an ultrasonic examination. However, though ultrasonic examination can give an indication of the percentage of the area which is bonded, it cannot provide any assessment of the quality of the bond. Another problem is that the welding parameters required to provide an adequate bond for a small area, such as a tube to tubeplate weld, may be entirely different to those required in large plate cladding. There may also be additional factors such as the state of contamination of the surfaces to be welded as might occur, for example, in the explosive plugging of liquid metal wetted surfaces in a liquid metal heat exchanger. Even if the weld satisfies both the definitions of an acceptable weld it might be that the explosive charge is excessive, and the product is unacceptable because of damage to the surface or because the product is badly distorted. It must be concluded that in experimental work aimed to investigate the parameters for an acceptable weld the assessment of what is and what is not acceptable is bound to be rather subjective.

The fundamental assumption made in the jetting theory discussed in Chapter 2 is that the metals at the point of collision behave as fluids, as the pressure at the collision point greatly exceeds the shear strength of the metals. Clearly if the impact velocity is too low then the metals at the collision point will not behave as fluids, and jetting will not take place. It can be concluded that there is a requirement for some minimum value of impact velocity in order to achieve an acceptable weld. Even if fluid-like behaviour is induced, it is still necessary that the thickness of the surface layers removed by the jet should be sufficient to completely remove the oxide, adsorbed gas and contaminant layer on the surfaces of both components. As the jet thickness depends on the dynamic angle of obliquity, β, it can be concluded that there is a minimum value of β to achieve an acceptable weld, which will be to some extent dependant on the surface preparation prior to welding. Surface roughness will also be of importance, as according to Shribman et al (88) excessive roughness can lead to an irregular weld interface with melted zones and porosity. This suggests that the thickness of material removed from the surface by the jet, should be appreciably greater than the surface roughness. Excessive surface roughness also makes it difficult or impossible to adequately clean the surface.

If the surface roughness is comparable to the thickness

of the material removed, which is certain to occur if the angle of obliquity is sufficiently small, then a thin cast intermetallic layer may be found at the interface, and if it forms a hard and brittle intermetallic compound it may be weakening. In well formed waves between explosively welded metals of similar density there will normally be vortices before and after the wave, or if the densities of the two metals are sufficiently different then only a single vortex before or after the weld will be formed. A considerable dissipation of energy occurs in the vortices as the kinetic energy of the jet is trapped in these areas, and according to Lucas et al (89) it is probable that there is always melting in the centre of the vortices. As the kinetic energy of the flyer plate is increased the molten zones at the centre of the vortices will grow in size and cooling cavities are observed. Ultimately, a level of kinetic energy will be reached when the molten pockets will coalesce to form a continuous melted zone. This can give rise to two problems; firstly, with some metal combinations brittle intermetallic compounds may be formed as between titanium and iron. Such welds will be unacceptable because of their low impact resistance. Secondly if the melt zone is too thick, then it may not cool sufficiently to develop an adequate hot strength before a reflected tension wave arrives which will pull the weld apart. All this suggests that there is an upper limit of the kinetic energy to produce an acceptable weld.

Proposals for the parameters to be satisfied to produce an acceptable weld have been made, amongst others, by Sedikh and Bondar (90), Carpenter et al (91), Kowalick and Hay (92) (93), Crossland and Williams (94), Wylie et al (95), Wittman (96), Carpenter and Wittman (97), Loyer et al (98) (99), Deribas et al (100), Stivers and Wittman (101), McKee and Crossland (26), and Takizawa and Izuma (102). Despite all the effort so far expended it has proved impossible to completely predict the conditions to be satisfied to achieve an acceptable weld, but nevertheless a large measure of understanding has been achieved. This can provide some guidance to those involved in applying the process.

In some early work Sedikh and Bondar (90) assumed that

interfacial waves were essential for adequately strong welds, as assessed by a simple shear test. Using an inclined set-up as depicted in Fig. 2.1 they concluded on the basis of experiments with steel/steel, copper/copper, copper/steel, and duraluminium/duraluminium welds, that a minimum flyer plate velocity of 1.5 km/s and a minimum initial angle of inclination of 1 - 2° are required to produce a weld. They pointed out, on the assumption of normal impact, that this minimum velocity was equivalent to a pressure of 30 GPa which they assumed was required to produce fluid-like behaviour in the surface layers. In the light of work since that time the assumption that interfacial waves were essential for adequate strength is incorrect, and the interpretation of their experiments was inadequate. The critical velocity quoted is very much higher than is strictly required. This is probably due to the fact that the collision point velocity is supersonic as the detonation velocity of the explosive used was 6 km/s, and consequently the critical value of the dynamic angle of obliquity, β_C, shown in Fig. 2.10 must be exceeded. If they had used a lower detonation velocity explosive then they would have found that adequate welds would have been produced at a much lower flyer plate velocity.

Carpenter et al (91), who used a nitrostarch-ammonium nitrate explosive with a detonation velocity of 3 km/s, limited their consideration to the parallel arrangement shown in Fig. 5.1, which is generally used for large plate cladding. They

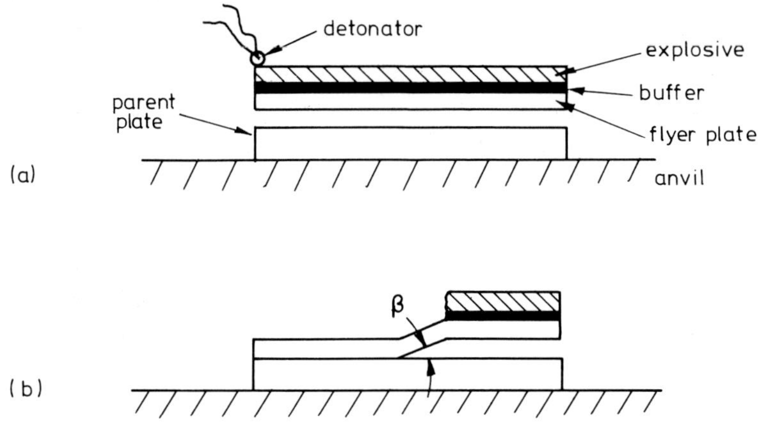

FIG. 5.1 Parallel set-up for explosive welding

concluded that the following parameters had to be satisfied.

(1) The collision point velocity, which in this situation is equal to the detonation velocity, must be less than the sonic velocity.

(2) The ductility of the flyer plate material should exceed 5% in a tensile test, in order to allow the dynamic bending indicated in Fig. 5.1.

(3) Solid solubility is important if melting takes place, but melting can be restricted to isolated pockets along the interface by the correct choice of welding conditions.

(4) From energy considerations coupled with experimental work involving measurement of the dynamic bend angle, they deduced a semi-empirical expression for the explosive loading:

$$L \propto \frac{\sigma \rho t}{d} \beta^2 \qquad (5.1)$$

where σ is the yield stress
L is the explosive loading or explosive mass per unit area
t is the flyer plate thickness
ρ is the density of the flyer plate
d is the stand-off distance
and β is the angle through which the plate is bent as shown in Fig. 5.1.

For a given value of $\frac{\sigma \rho t}{d}$ they suggest there is a minimum value of explosive loading to give an acceptable weld, and a maximum value above which melting occurs.

Equation 5.1 is dimensionally non-homogeneous and the graph given by the authors uses a non-coherent system of units. More importantly, the derivation of the equation is based on the assumption that the collision parameters can be changed by varying the stand-off distance, d. For the explosive they were using, it is probable that the terminal velocity of the flyer plate is reached within a stand-off distance equal to the flyer plate thickness, that is $d = t$, so their assumption does not have general validity. There is also the question what value of the yield stress should be taken as the process involves very high and ill defined strain rates. Nevertheless, their suggestions were a marked step forward.

Kowalick and Hay (92) suggested that a good weld is associated with regular wave formation at the interface, and this is a commonly held belief. However, McKee and Crossland (26) then reported on copper to copper welds achieved with a high velocity gas gun in which they determined the removal of material by weighing, which is some indication of jetting. They found for a fixed velocity and varying the impact angle, β, that the maximum removal of material occurred at the angle at the borderline between a wavy and straight interface. This perhaps suggests that an equally strong bond would be achieved either side of this borderline. The belief that interfacial waves are desirable probably derives from the fact that with the explosives readily available, and in order to satisfy the more obvious welding parameters, the conditions for wave generation are also satisfied. Kowalick and Hay noted from the results reported by Bahrani and Crossland (39) that there appears to be a transition from laminar to turbulent waves in the experiments reported on welding a flyer plate to a semi-cylindrical parent plate. They considered that for an acceptable weld turbulent waves should be avoided because of the danger of formation of intermetallics in the vortices. Using dimensional analysis they derived an equation for the explosive mass per unit area at this transition from streamline to turbulent waves:

$$L = k^1 \rho t C \beta^2 \qquad (5.2)$$

where k^1 is a constant, C is the acoustic velocity and the rest of the symbols have the same meaning as in Equation 5.1. They were not explicit which acoustical velocity they were referring to or for which of the component materials in a weld between dissimilar metals. From experimental work on welding a stainless steel flyer plate to a mild steel parent plate using low-density nitroguanidine explosive with a density of 400 kg/m^3, they deduced a value of k^1 of $(123 \pm 16.7) \times 10^{-5}$ s^2/m^2, where the other quantities are in SI units, for the transition from streamline to turbulent waves. They suggested that a range of

k^1 would describe the weldability window, but they did not appear to pursue this approach any further. Even if Equation 5.2 adequately describes the transition from streamline to turbulent waves, this does not seem to be a physical justification for using this form of equation to describe the weldability window.

Crossland and Williams (94) reviewed the literature on welding parameters, and they suggested parameters for an accept-weld which were the basis of an experimental programme reported by Wylie et al (95). Wylie et al suggested the following parameters:

(1) In order to avoid the limitations on jetting under supersonic conditions which have been noted in Chapter 3, V_W/C_1 and V_F/C_2 should be less than unity, where C_1 is the sonic velocity in the parent plate material and C_2 the velocity in the flyer plate material.

(2) The impact angle, β, should exceed a limiting value below which welding will not occur regardless of V_p. This value of β may be related to the thickness of the contaminant surface film.

(3) A minimum value of contact pressure, which is related to V_p, must be achieved to produce fluid like behaviour at the interface.

(4) Above this minimum impact velocity the conditions for an acceptable weld appear to be associated with the kinetic energy imparted to the flyer plate by the explosive charge. The kinetic energy for a satisfactory weld may be related to the strain energy to cause dynamic yield, and it also appears to be related to the stronger of the two materials.

(5) For metal combinations where a brittle intermetallic compound can be formed there is an upper limit of the impact energy beyond which the weld is weakened by the formation of excessive intermetallic compound. Even where brittle intermetallic compounds are not a problem there will still be an upper limit, because if there is

excessive melting at the weld interface then the weld created may be torn apart by the reflected tension wave from one of the free surfaces, before the metal at the weld interface has developed sufficient strength.

(6) The stand-off or clearance between the flyer and parent plate in a parallel welding arrangement, as shown in Fig. 5.1, should exceed half the flyer-plate thickness to allow the flyer-plate to achieve its maximum velocity.

(7) In general, the use of high detonation-velocity explosives should be avoided as the very high pressure pulse produced in the flyer-plate can, while the flyer-plate is in flight, produce a reflected tension wave which can in turn lead to spalling of the flyer-plate.

(8) There appears to be no effect of anvil characteristics on the weld quality, but in practice this could well affect the damage done to the product and in particular the amount of distortion of the product.

These parameters satisfy many of the limitations found by experience, but they rely on experience to define the numerical values for the minimum flyer-plate velocity and for kinetic energy, the minimum value of the dynamic angle of obliquity and the upper limit of kinetic energy. However, they do not completely define the boundary within which welding can be achieved in the $^{V}p/_{\beta}$ plane, which requires some parameter such as an upper limit for β. From what has been written in Chapter 4 it would appear that the stand-off to allow the flyer-plate to reach its terminal velocity should be equal to the flyer plate thickness for explosives such as Trimonite 1, and twice the flyer-plate thickness for Anfo. Lastly, it appears unlikely that anvil characteristics have no effect on the quality of the weld. For instance if the parent plate was lightly suspended or if it was resting on some very soft material, then as indicated in Fig. 5.2 the angle of obliquity would appear to be reduced due to the momentum transferred to the portion of the parent plate which has been impacted by the flyer plate. If at

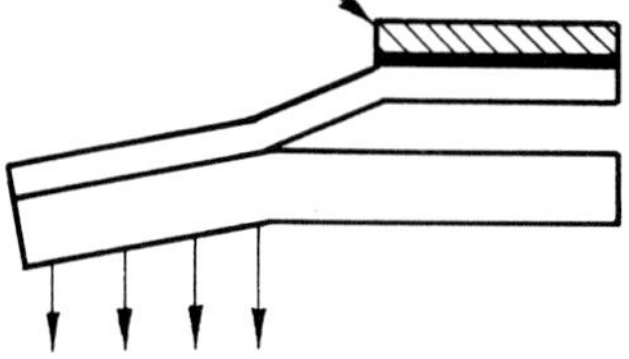

FIG. 5.2 Effect of lightly supported parent plate

the other extreme the parent plate was supported on a massive anvil plate, then the angle of obliquity would not be affected by momentum transfer to the parent plate. Obviously the effect of momentum transfer will be much less if the flyer plate is much thinner than the parent plate which is normally the case.

Crossland (103) using the parameters suggested by Wylie et al (95) has examined the limitations on welding brass to mild steel. For these calculations he took the minimum values of the flyer-plate velocity, kinetic energy and dynamic angle of obliquity as 200 m/s, 150 J/cm^2 and 7° respectively, and the maximum kinetic energy as 450 J/cm^2. For a parallel set-up such as shown in Fig. 5.1 the angle β can be determined from Equation 2.14 if V_P and V_D are known. Table 5.1 gives the essential information using Trimonite 1 explosive, as the thickness of the flyer-plate is increased; the necessary data for Trimonite 1 is given in Chapters 3 and 4. Up to a flyer plate thickness of 8 mm the kinetic energy is kept at the minimum value, but above 8 mm the kinetic energy must be raised in order that the minimum value of V_P is maintained. It will be seen that even for a 4 mm thick plate the angle β is slightly less than that specified, so that nominally it would not be possible to exceed 4 mm, though it should be noted that it critically depends on what is the permissible lower value of β.

To weld thicker flyer-plates it is possible to increase the explosive loading and hence V_P and β, but if the maximum kinetic energy is exceeded then the weld quality deteriorates and excessive damage to the product may occur. Table 5.2 gives data aimed to establish the maximum flyer-plate thickness which might be welded using Trimonite 1 explosive. It

TABLE 5.1

Flyer-plate Thickness t mm	Flyer-plate Velocity V_P m/s	Kinetic Energy J/cm^2	Explosive Loading R	Thickness of Explosive mm	Detonation Velocity V_D km/s	β^o
2	400	150	0.77	11	2.61	9.7
4	316	150	0.4	11.4	2.63	6.8
6	250	150	0.24	12.6	2.69	5.4
8	218	150	0.22	14.2	2.81	4.4
10	200	158	0.21	15.0	2.88	4.0
12	200	190	0.21	18.1	3.04	3.8
16	200	252	0.21	24.1	3.23	3.6
20	200	316	0.21	30.1	3.38	3.4
24	200	380	0.21	36.1	3.50	3.3
28	200	442	0.21	42.0	3.60	3.2

TABLE 5.2

Flyer-plate Thickness t mm	Flyer-plate Velocity V_P m/s	Kinetic Energy J/cm^2	Explosive Loading R	Thickness of Explosive mm	Detonation Velocity V_D km/s	β^o
10	310	380	0.4	28.5	3.37	5.3
10	375	555	0.5	36.0	3.6	6.0
10	418	685	0.6	43.0	3.61	6.6
10	450	800	0.7	50.0	3.7	7.0
8	310	305	0.4	22.7	3.21	5.6
8	375	445	0.5	28.8	3.37	6.4
8	418	550	0.6	24.5	3.48	6.9
8	450	640	0.7	40.0	3.58	7.2
6	310	227	0.4	17.1	3.03	5.9
6	375	333	0.5	21.5	3.18	6.8
6	418	410	0.6	25.8	3.29	7.3
6	450	480	0.7	30.0	3.38	7.7

will be seen that it would be possible to weld 6 mm thick flyer-plate, and satisfy the limitation imposed, and if the minimum value of β could be reduced by only 1° to 6° then 8 mm thick flyer-plate could be welded.

If it is necessary to weld even thicker flyer-plate then the only practical proposition would be to use a lower detonation velocity explosive such as Anfo, details of which are given in Chapters 3 and 4. Table 5.3 has been calculated for Anfo; for the thicknesses of the explosive layers required it can be assumed that the detonation velocity is 2.5 km/s. From Table 5.3, which is based on the upper bound of kinetic energy, it will be seen that it is possible to weld flyer-plates up to 12 mm thickness or even 16 mm if the minimum value of β could be reduced to 6°.

TABLE 5.3

Flyer-plate Thickness t mm	Flyer-plate Velocity V_p m/s	β°	Flyer-plate Thickness t mm	Flyer-plate Velocity V_p m/s	β°
10	340	7.8	20	240	5.6
12	310	7.2	24	228	5.3
16	369	6.2	28	202	4.7

Using these welding parameters it is possible to examine the lower limit of flyer-plate thickness, which can be welded. With Trimonite 1 explosive it is found that for a flyer-plate thickness of 0.5 mm, the angle β is probably unacceptably large. This might be overcome by using an explosive with a detonation velocity closer to the velocity of sound, which might then allow the welding of a 0.2 mm thick flyer-plate. Even thinner flyer-plates might be welded if they were stuck to a driver-plate, that is a plate which is between the flyer-plate and the explosive, and which is prevented from welding to the flyer-plate by the adhesive layer or some other non-metallic film between them. The driver-plate effectively increases the thickness of the flyer plate, though at an appreciable additional cost.

The parameters proposed by Wylie et al (95) were a marked step forward in understanding the parameters which had to be taken into account in explosive welding. Unfortunately, it is difficult, if not impossible, to design explosive welding experiments to allow a parametric study of the proposed parameters. Despite extensive experimental work it proved impossible to closely define the minimum values of impact velocity, kinetic energy, and dynamic angle of obliquity and the maximum value of kinetic energy. It is also probable that the state of surface cleanness and roughness are important in defining these quantities. To overcome this difficulty of carrying out a parametric study McKee and Crossland (26) and Szecket and Crossland (104) have developed two gas guns, which allow a relatively small flyer-plate to be launched at a controllable velocity against an inclined target plate. As an alternative, Meyer (105) used electro-magnetic propulsion of a flyer-plate. These workers all examined the welding of half hard copper to half hard copper, but Szecket (106) alone has obtained data which completely delineates the boundaries of the zone in which welding occurs or what is now referred to as the explosive welding window. These results are shown in Fig. 5.3. It will

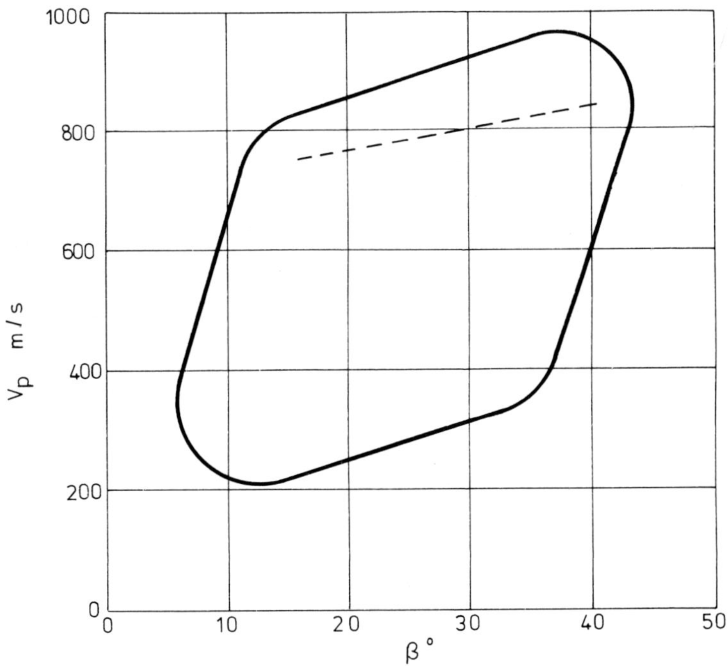

FIG. 5.3 Welding window for half hard copper to half hard copper (106). —— 1.58 mm thick flyer plate, ---- upper limit for 3.17 mm thick flyer plate

be noted that the boundaries of the window are not perfectly in accord with the suggestions made by Wylie et al (95). For instance the lower boundary is not at a constant value of V_p and the left hand boundary is not at a constant critical value of β. The top boundary for this combination is ill defined, as though welding occurs at even higher velocities the flyer-plate is considerably damaged and reduced in thickness. Such welds are clearly unacceptable though the bond itself appears adequate. Under these circumstances what is or what is not an acceptable weld is a purely subjective judgement.

The essential problem of the parameters proposed by Wylie et al (95) is in the determination of the limiting values; it would be desirable to be able to either calculate these values or determine them by a limited number of experiments. Wittman (96) and Deribas et al (100) have used a plot of V_W/β to describe the welding window. Fig. 5.4 shows the boundaries

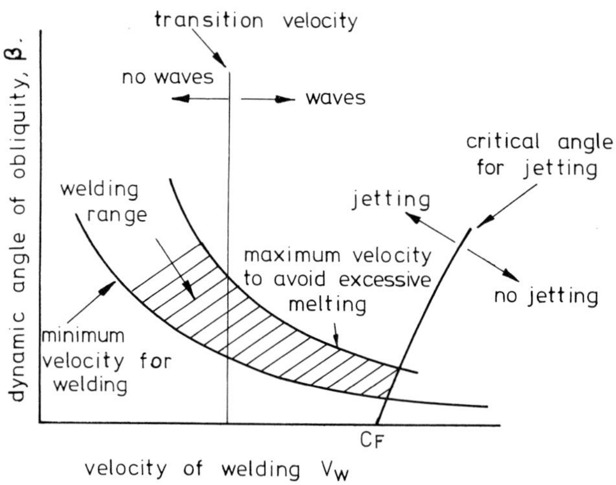

FIG. 5.4 Welding window proposed by Wittman (96) and Deribas et al (100)

of this welding window. On the right there is a boundary in the supersonic region where jetting cannot occur unless the angle of obliquity exceeds a critical angle, as described by Walsh et al (27) and Cowan and Holtzman (24) and discussed in Chapter 2. The lower boundary defines the conditions to achieve fluid-like conditions at the interface so that jetting occurs. The upper curve is the limit above which excessive defects occur at the weld interface, due to the hot tensile strength being insufficient to withstand the reflected tensile wave. Another feature which can be represented is the flow transition between wavy and straight interfaces. No boundary to the left-hand side of the window has been defined.

Deribas et al and Wittman appear to agree that the lower limit is given by the relationship:

$$\sin \beta = k_1 \sqrt{\frac{H_V}{\rho V_W^2}} \tag{5.3}$$

where k_1 is a constant, H_V is the Vickers hardness in N/m^2 and ρ is the density in kg/m^3. However, Deribas et al note that Equation 5.3 gives values which differ from experiment at large values of β, so they suggest that Equation 5.3 should be written as:

$$\beta = k_1 \sqrt{\frac{H_V}{\rho V_W^2}} \tag{5.4}$$

where β is in radians. The value of k_1 is 0.6 for a high quality pre-cleaning of the surfaces, and 1.2 for less perfectly cleaned surfaces; further work needs to be done on the effects of surface contamination and roughness.

It is not clear from either of these papers if the hardness, H_V, is that of the flyer or parent plate, but according to Loyer et al (99) it is only necessary to satisfy the jetting conditions for the softer of the two materials. The static hardness is used, nevertheless it is a dynamic situation and it must be questioned if this can be justified.

The approach of Deribas et al and Wittman to defining the upper limit is very different; Wittman determines the value of V_P above which interfacial melting will occur, while Deribas et al accept that melting occurs and determine the condition that the time for solidification should be less than the time for the tensile wave to reach the interface. This would suggest that the Wittman analysis should give a lower bound solution, while the analysis due to Deribas et al should give an upper bound. Both solutions lead to complex equations involving a lot of physical constants, but as they both involve unknown empirical coefficients it is adequate to clump together the physical constants and empirical coefficients. If this is done then Wittman's equation reduces to:

$$V_P = \frac{k_2}{t^{\frac{1}{4}} V_W} \tag{5.5}$$

where k_2 is a constant and t is the flyer-plate thickness. From V_P and V_W it is possible to determine β from Equation 2.11 or 2.14, depending if it is an inclined or parallel plate arrangement. The equation due to Deribas et al becomes:-

$$\sin\frac{\beta}{2} = \frac{k_3}{t^{\frac{1}{4}} V_W^{1.25}} \tag{5.6}$$

The transition velocity between a wavy and a straight interface is given by Equation 2.19.

The data obtained by Szecket (106) for welds between half hard copper and half hard copper is plotted in Fig. 5.3 on the V_W/β plane, and it is replotted in Fig. 5.5 on the β/V_W plane. It can be seen that the lower limit is well fitted by Equation 5.4 if the value of k_1 is taken as 0.85, which lies within the range of 0.6 for high quality pre-cleaning of the surfaces and 1.2 for less perfectly cleaned surfaces. For the upper limit Wittman (96) suggests that the constant k_2 in his Equation 5.5 should be evaluated experimentally at a value of V_W which is equal to half the compressive wave velocity, C_F, given by Equation 4.6, which for copper is about 1970 m/s. It will be seen that the resultant curve does not fit the experimentally determined upper limit. Using Equation 5.6 proposed by Deribas et al (100) and determining the constant k_3 at the same value of V_W it will be seen that there is good agreement with the experimentally determined upper limit. There is also good agreement with the experimental upper limit for a flyer plate of twice the thickness, data for which is given in Fig. 5.3, if the same value of k_3 is used. The right hand limit does not appear to be provided by the conditions for jetting in the supersonic region, but it appears to be limited to the compressive wave velocity, C_F, given by Equation 4.6 or in Table 4.2. This particular area of the explosive welding window is very sensitive to the precise boundary at low values of β in the V_P/β plane shown in Fig. 5.3, so the experimental data is not

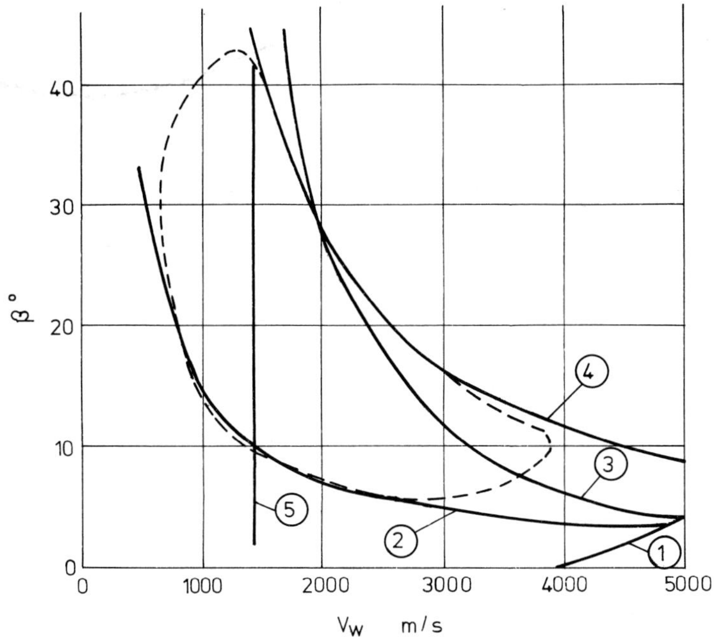

FIG. 5.5 Explosive welding window for 1.58 mm half hard copper flyer plate to half hard copper parent plate (106). ---- Experimental welding window, (1) limits of jetting in the supersonic region, (2) lower limit curve for welding due to Deribas et al (100), (3) upper limit curve due to Wittman (96), (4) upper limit curve due to Deribas et al (100) and (5) transition between smooth and wavy flow.

sufficiently discriminating to investigate the validity of the proposed boundary limitation in this region.

Neither Wittman (96) or Deribas et al (100) proposed a boundary condition at the left hand side of Fig. 5.5, but Loyer et al (98) (99) have suggested without a clear physical explanation that there is a lower limit for V_W if fluid like conditions are to be satisfied. It will be seen that this is not satisfactory unless a value of V_W is assumed which is equal to the transition velocity between smooth and wavy flow, given by Equation 2.19. However, if this value is assumed it eliminates a large area of the window in which satisfactory welds are produced.

The data given by Szecket (106) enables the validity of the equations given by Cowan et al (35) for the transition velocity between smooth and wavy flow to be checked. Fig. 5.6

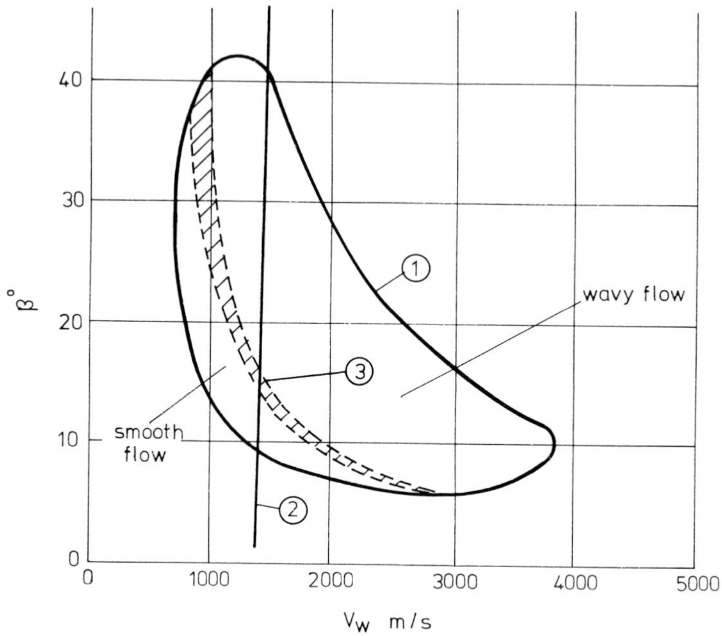

FIG. 5.6 Transition from smooth to wavy flow. (1) explosive welding window Szecket (106), (2) theoretical transition velocity Cowan et al (35) and (3) experimental transition region

shows the experimentally determined transition between smooth and wavy flow, and it will be seen that a constant value for this velocity is not substantiated, and in particular the value given by Equation 2.19.

Loyer et al (98) (99) have examined the explosive welding window, but they have chosen the R/V_W plane, where R is the explosive loading ratio or the ratio of mass of explosive to mass of the flyer plate. This lacks the generality of the V_P/β and the β/V_W plots used by other workers in the field, as it only applies to the particular explosive used and if another explosive is used then the explosive loading ratio will have to be adjusted. Their plot is shown diagrammatically in Fig. 5.7. Boundary 1 is the minimum flyer-plate velocity required to generate fluid-like conditions at the interface, and basically this is defined by Equation 5.4 in which they substitute:

$$\beta = \tan^{-1} \frac{V_P}{V_W} \simeq \frac{V_P}{V_W}$$

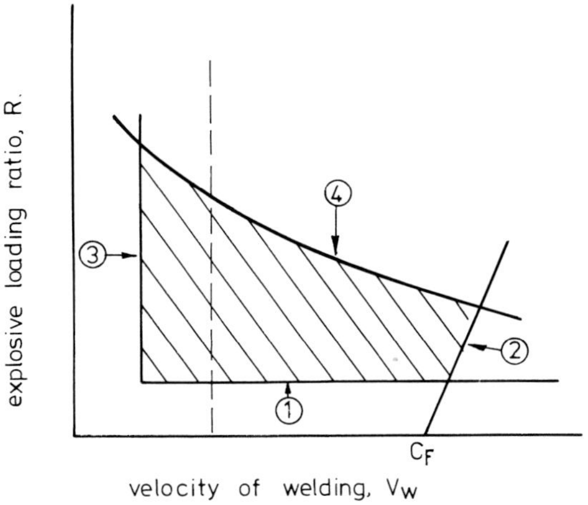

FIG. 5.7 Welding window proposed by Loyer et al (98) (99)

so Equation 5.4 reduces to:

$$V_P = k_1\sqrt{\frac{H_V}{\rho}} \quad (5.7)$$

where they have taken a value of 1.14 for k_1.

Boundary 2 defines the critical angle for a detached shock wave under supersonic flow as defined by Walsh et al (27) and Cowan and Holtzman (24). Boundary 3 is the minimum velocity of welding to produce flow near the collision surface, though it is not clear what is precisely meant by this statement. Lastly, boundary 4 is the upper limit of flyer-plate kinetic energy to avoid defects at the weld interface, due to the hot tensile stress being insufficient to withstand the reflected tension wave, for which they have used the analysis proposed by Wittman (96).

Loyer er al have explored the welding window for the following combinations Al/Al, Cu/Cu, Cu/Al, Steel/Steel, and Al/Steel, and for each of these they have produced micrographs round and within the periphery of the window. They reached several very important conclusions:

(1) The upper limit of the welding window cannot be adequately defined by a unique ratio of R, as it was recognised that it depends on the kinetic energy which is a function of flyer plate thickness. This is in agreement with Wylie et al

(95), Wittman (96) and Deribas et al (100).

(2) For Al/Steel the upper energy limit is close to that of the parent metals, while for Cu/Al it is probably less due to the formation of brittle intermetallics. This is in agreement with Wylie et al (95).

(3) The value of R at the lower limit of the welding window appears to be dictated by the softest of the component metals, if account is taken of the different acoustic impedance of the metals. This suggests that the jet formed by the yielding of only one of the metals, adequately cleans the surface of the other metal.

(4) Aluminium can be welded to steel but not steel to aluminium.

(5) It proved impossible to weld Al-Mg-Mn to steel or vice versa.

There seems to be general agreement about the nature of the welding window, and this can be used to indicate the possible limitations imposed in practice. For this purpose the welding window established for half hard copper to half hard copper by Szecket (106) and shown in Fig. 5.5 will be used. The lower boundary appears to be well fitted by Equation 5.4 if a value of k_1 of 0.85 is taken, which lies reasonably between 0.6 and 1.2 quoted by Deribas et al (100) and Wittman (96). In practice it might well be advisable to take a value of k_1 of 1.2 or even greater, to allow for the quality of the surface preparation of the flyer and parent plates which it is reasonable to maintain when preparing large plates, and required to provide a factor of safety to take account of the variability of the explosive. If a value of 1.2 is taken then this gives the lower curve of Figs. 5.8 or 5.9.

The upper curve in Fig. 5.5 was obtained for a 1.58 mm thick flyer plate, and it appears to fit the form of Equation 5.6. Consequently it is possible to deduce the upper boundary curve for other thicknesses of flyer plate; these are shown in Figs. 5.8 and 5.9. It is worth noting that the limited data for the upper boundary curve for a flyer plate of thickness 3.17 mm, which is shown in Fig. 5.3, appears to be in good

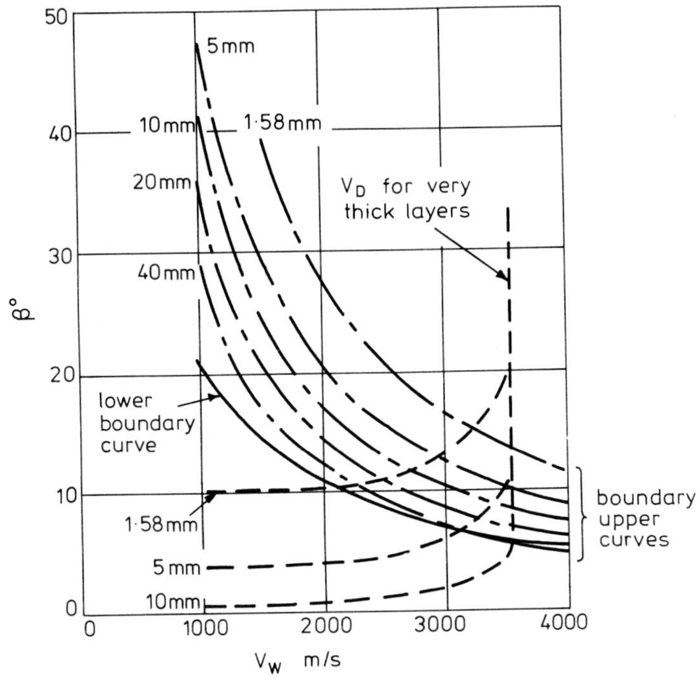

FIG. 5.8 Upper and lower boundary curves for half hard copper to half hard copper, with the curves which can be obtained using Trimonite 1 explosive

agreement with the values which would be predicted from the 1.58 mm flyer plate data.

It will be seen that the welding window becomes much narrower as the flyer plate thickness is increased. From the present data and using Equation 5.5 it would seem that the risk in welding a 40 mm thick flyer plate would probably not be acceptable. Even for a 20 mm thick flyer plate it would be desirable to work in the region of V_W less than 2000 m/s, where the width of the welding window is greater.

For large plate cladding the parallel set-up will generally be used and consequently V_D and V_W are equal. For an explosive whose properties are known it is possible to plot a family of curves for varying flyer plate thickness on the V_W/β plane. Assuming a value for the explosive loading, R, the flyer plate velocity, V_P, can be calculated from Equation 4.12. Knowing the explosive loading and flyer plate thickness the thickness of the explosive can be deduced, and then from data such as that given in Figs. 3.10 and 3.11 the detonation

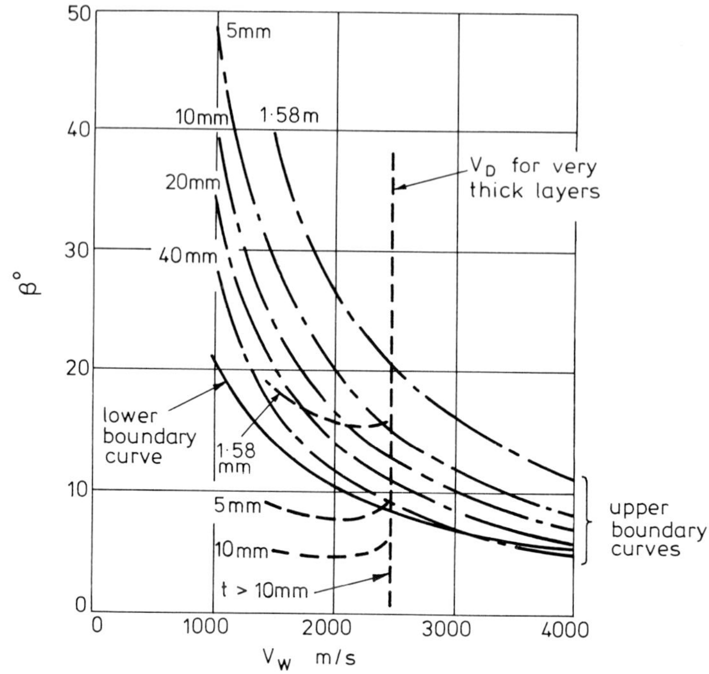

FIG. 5.9 Upper and lower boundary curves for half hard copper to half hard copper, with the curves which can be obtained using 6% Anfo explosives

velocity, $V_D (= V_W)$, can be found. Using Equation 2.14 and knowing V_P and V_D the value of β can be determined. The family of curves for Trimonite 1 are given in Fig. 5.8 and for Anfo (6% diesel) in Fig. 5.9.

It will be seen in Fig. 5.8 that it would only be possible with Trimonite 1 explosive to satisfactorily weld thin flyer plate. Even with a flyer plate of 5 mm thickness the operating region is very small, and it will be seen that it is right in the toe of the welding window shown in Fig. 5.5, where the theoretical upper and lower bounds overestimate the boundaries of the welding window. This suggests that Trimonite 1 explosive is not an acceptable explosive for this particular operation, except for very thin flyer plates. Fig. 5.9 shows a similar family of curves for Anfo (6% diesel) explosive, and it will be seen that the operating region is much improved and it should be possible to weld up to a 20 mm thick flyer plate or even greater. The situation might be further improved by using an explosive with an even lower maximum detonation velocity. This could be achieved by adding inert material, such as salt, to the explosive. From Fig. 5.6 it will be seen that the operating regions for both Trimonite 1 and Anfo explosives are in the area where waves are generated, but equally acceptable welds with no waves would be possible with an explosive with a lower detonation velocity.

Though the physical nature of the boundary conditions within which welding occurs seems to be largely agreed, there still remains much which has not been adequately explored. For instance it does not appear to be reasonable for the static hardness to be used in Equations 5.4 and 5.7, in what is a dynamic situation, where some materials are more strain rate sensitive than others. There is also the question of which hardness to use when welding component metals of different hardnesses? Is it satisfactory to use the hardness of the softer material as suggested by Loyer et al (99)? The effect of surface contamination and roughness on the lower limit has recently been reported on by Dobrushin (107), but further work is required to assess what values should be assumed for the constant k_1 in Equation 5.4. Is 1.2 adequate or should a higher value be taken in order that the quality of surface finish required is not prohibitively expensive? At the upper boundary the generality of Equation 5.6 has not been established, nor has the validity of the more detailed analysis proposed by Deribas et al (100) been explored. The Deribas et al analysis at the upper boundary is not physically adequate, as the real limit is not that the molten zone at the interface shall have solidified by the time the tensile wave reaches it, but that the solidified zone should have developed adequate hot strength to withstand the stress wave. In welds between different metals the problem is even more complex as there is a complete lack of information on the hot strength of the intermetallic compounds or mixtures at the interface. Lastly there is still no understanding of the physical nature of the left hand boundary in the β/V_W plane.

What is required by the practioner of explosive welding is either an adequate theoretical description of the boundaries of the explosive welding window for the combinations he is interested in, or a compendium of experimentally determined explosive welding windows. This would allow the operator to decide on welding conditions to provide a satisfactory weld. Though a gas gun such as that used by Szecket and Crossland (104) may allow a rapid determination of the window, it does not provide a sufficiently large specimen to be truly representative, and it needs to be supplemented by explosive welding trials. A simple and economical explosive test is provided by welding a flyer plate to a semi cylindrical parent plate as used by Bahrani and Crossland (39) and Kowalick and Hay (93), though a fairly large semi-cylindrical plate with a diameter of 100 mm or more is desirable, and of sufficient width to give a reasonable aspect ratio. It is an economical test as each specimen provides a mass of data, as though the flyer plate velocity is constant the dynamic angle of obliquity, β, varies as the flyer plate wraps itself round the semi-cylindrical parent plate.

6

FLAT PLATE CLADDING

There is an increasing demand for cladded plates, but up till the development of explosive welding it was largely an unsatisfied demand, as it was only possible to produce clad plate by roll cladding or weld overlay. Hot roll cladding is essentially a pressure weld, achieved by passing a carefully prepared sandwich through a rolling mill. This requires a lot of preparation of the sandwich and it requires capital intensive plant. As has been shown by Vaidyanath et al (108), considerable deformation is required in roll cladding to achieve an adequate bond strength, and many material combinations are not amenable to roll cladding. With weld overlay the cost of covering a large area is considerable, and the combinations which can be produced are limited to those which are compatible under fusion welding. For instance it is not possible to weld overlay metal combinations, the components of which have very different melting points or which form brittle intermetallics when welded together. The main product of roll cladding and weld overlay has been stainless steel cladded to steel, but there are many other combinations of interest to industry particularly the chemical industry.

The reasons for using composite plate instead of solid material is the need for corrosion resistance at lower cost, improved heat transfer characteristics, acceptable strength or stiffness at reduced cost, improved electrical properties, or better abrasion or erosion resistance. For some chemical processes it is necessary in order to prevent chemical attack or to improve heat transfer, to use very expensive metals for the containing vessels such as nickel, hastelloy, stainless steel, titanium or tantalum. For example tantalum has an outstanding resistance to acid chemical attack, however, its utilization in large chemical process equipment is limited by the availability of large sheets and more particularly by cost. The material cost alone for a large vessel made from such materials would be prohibitively expensive. In the fabrication of such vessels there is a plate thickness beyond which it is economic to use explosively cladded plate, even allowing for the additional cost in making welded joints between composite plates. Some very large vessels have been fabricated from explosively cladded plate, for instance the Nooter Corporation of St. Louis, Mo., have used titanium/steel cladded plates for vessels used in the manufacture of aromatic acids. One such vessel, for example, is 2.44 m diameter by 7 m long, and it was fabricated from plate consisting of 2 mm thick titanium cladded to 22 mm thick SA-516 grade 70 steel. They have also produced a vessel 2.44 m diameter by 18.3 m long fabricated from tantulum/copper/steel explosively cladded plate.

McKenny and Banker (109) have demonstrated the application of aluminium/steel explosively welded transition joints to ship construction, and in particular the welded connection between the steel hull and the aluminium superstructure. Schetky (110) has discussed the potential economic advantages of constructing large and small ships from cupro-nickel cladded steel plate, which arises from the anti-fouling properties of cupro-nickel. The economic advantage accrues from reduced fuel consumption, higher operating speeds, reduction in the frequency of dry-docking for hull maintenance, no need to paint the hull and lack of hull corrosion.

The low electrical resistance and resulting low power loss of explosively welded transition joints is an advantage in electrolytic processes. Enright et al (111) have discussed the successful application of aluminium/steel and aluminium/copper explosively cladded plates, in the fabrication of anodes for aluminium reduction cells. Persson (112) has reported on the application of titanium/silver/steel explosively clad plate to electro-chemical cells. The tri-clad, which consists of 12 mm titanium, 0.4 mm silver and 22 mm steel, is produced in two separate cladding operations, and thin titanium plates are then fusion welded to the titanium side of the tri-clad, and similarly steel to the steel side.

Takizawa (113) has noted that the Japan National Railways have constructed a computer controlled marshalling yard, with remote controlled shunting locomotives using linear motors which run between the normal rails. The reaction plate of the linear motor is laid between the rails, and it consists of aluminium/steel clad plate, with 3 mm of aluminium bonded to the 16 mm of steel, the plates being 230 mm wide. This has an excellent combination of desirable electrical properties, strength and rigidity.

In heat exchangers the tube plates not only require to be made of corrosive resistant metals, but they also require to be adequately stiff, so that the deflection due to a pressure difference across the plate may be kept small. Materials such as naval brass and monel metal are frequently used for tube plates, and to increase the rigidity very thick plates may be required as they are low modulus materials, but the cost of large thick tube plates of brass or monel metal will be considerable. Alternatively the stiffness of the tube plate can be increased by using monel metal or naval brass cladded to steel, which is a higher modulus material and also cheaper.

Armour plate is yet another potential application for explosively clad plate. Manganello and Abbott (114) have pointed out that the resistance of steel targets to ballistic penetration may be increased, by bonding a very hard front sheet to conventional armour plate, though they add that the improved performance is very dependant on a strong metallurgical bond at the interface. Basically the hard front surface breaks up the armour piercing shot, but as the fracture toughness of the hard material is likely to be low, it would normally disintegrate unless it was bonded to a high fracture toughness back plate. Crossland et al (115) have reported on experiments in which they explosively welded together steels of varying hardness, for instance they have welded 6.4 mm thick maraging steel plate in either the annealed or heat treated state to 25 mm thick armour plate steel. In the case of the annealed maraging steel plate welded to armour plate, the product was subsequently heat treated at 490°C, which age hardened the maraging steel while subjecting the armour plate steel to a mild stress relieving heat treatment. The resulting product had a hardness in the maraging steel of 530 VPN with probably a low impact strength, and in the armour plate steel back plate the hardness was 298 VPN with a high impact strength.

These few examples indicate the potential practical value of explosively clad plate, but it does not indicate the large range of possible combinations of metals which can be produced. The inherent versatility of explosive welding permits the production of a wide variety of clad plate products. It appears that most metals can be explosively cladded together, as long as they have adequate ductility to tolerate the deformation imposed in the process. There will be limitations on the thickness of the flyer plate imposed by the welding window. However, the most frequently exploited combinations are of various metals cladded to steel, and in particular stainless steel, copper, brass, naval brass, aluminium bronze, nickel Monel, cupro-nickel, Hastelloy alloy, incoloy alloys, inconel alloys, silver, tantalum, titanium and aluminium. Other combinations include aluminium bronze/stainless steel, titanium/stainless steel, titanium/copper, copper/aluminium, stainless steel/steel/stainless steel, titanium/steel/titanium, tantalum/copper/steel, aluminium/titanium/steel, titanium/stainless steel/steel, aluminium/stainless steel/steel, and aluminium alloy/aluminium/titanium/nickel/stainless steel. The thickness of the cladding material depends on the particular metal, the application and the size of plate to be clad, but typically it will vary from less than a millimetre up to ten or twenty millimetre and in some cases even higher. The minimum thickness of the cladded plate is normally twice the thickness of the flyer plate, but usually it is much thicker and there is no upper limit imposed by the process. The area of plate which can be cladded depends on the size of cladding and base materials available, though in some cases two plates can be fusion welded together to form a larger plate. The size of the charge will depend on the size and thickness of the flyer plate, and it will be limited by the site available. For example, the Nobel's Explosives Co. Ltd. on their explosive site in New Galloway in Scotland can clad shell plates for pressure vessels up to areas of 12 m^2 and tube plates up to 3.5 m diameter. Persson (116) of Nitro Nobel in Sweden states that for 12 mm thick stainless steel cladded to steel the largest plate they can clad is 6 m by 2.5 m, and they can clad even thicker plates up to 45 mm but they would then be restricted to smaller areas. The Asahi Chemical Industry Ltd. in Japan have a large man-made chamber similar to that shown in Fig. 3.15 at their Aibano site and an even larger tunnel at their Chikushina site. They claim that they can clad an area of 2.5 m by 8 m with a 2 mm thick flyer plate or 1.45 m by 4.6 m with a 13 mm thick flyer plate. Figures 6.1 - 6.8 show a few examples of flat plate cladding.

FIG. 6.1 Ten charges with a total mass of 7000 kg being detonated simultaneously (Explosive Fabricators Inc.)

FIG. 6.2 Laying and levelling the explosive (Explosive Fabricators Inc.)

FIG. 6.3 A 3.2 m diameter aluminium bronze (15 mm)/steel (115 mm) tubeplate, and the heat exchangers which incorporated it (Nobel's Explosive Co. Ltd. and Westgarth Ltd.)

FIG. 6.4 A stainless steel (21 mm)/steel slab for hot rolling into stainless steel clad sheet (Nobel's Explosives Co. Ltd.)

FIG. 6.5 A tantalum clad column fabricated by the Nooter Corp., St. Louis, Mo. (Dupont Detaclad)

FIG. 6.6 Vessels fabricated by Construzioni Meccaniche, Milan Italy from 2 mm titanium clad to 21 mm 5A-516 grade 70 steel (Dupont Detaclad)

FIG. 6.7 Entrance to explosive welding chamber at the Aibano Works of the Asahi Chemical Industry Co. Ltd.

FIG. 6.8 Marshalling yard using linear motors with an aluminium (3mm)/steel (16 mm) clad reaction plate (Asahi Chemical Industry Co. Ltd.)

Flat plate cladding has been the subject of numerous patents, which tend to obscure rather than clarify the essential practical problems in operating the process. Many papers have been published on small scale experiments using thin flyer plates, in which their authors have been concerned with fundamental research into the mechanism and metallurgy of explosive welding, rather than resolving the practical problems associated with large plate cladding. Ignoring the process patented by Phillipchuk et al (7), which was based on what is now accepted as a completely incorrect understanding of the mechanism, the first arrangement to be used was the symmetrical set-up shown in Fig. 2.5, which was described by Pearson (10). However, it would be difficult or impossible to apply it to large plates. For instance, if one plate was placed on the ground with its explosive charge underneath it, then the composite plate would be likely to be hurled into the air. Alternatively if the two plates were arranged vertically, then there is the problem of how to keep the explosive in intimate contact with the plates, especially with a powder explosive. Again there would be a difficulty in preventing the plate being fired horizontally at a reasonable velocity.

In general the commercial need is for plates, from less than one millimetre up to twenty or thirty millimetre thickness, to be cladded to a much thicker base plate, and the inclined process shown in Fig. 2.1 might appear ideal. This geometric arrangement allows a high detonation velocity sheet explosive to be used, which is more convenient to place than powder explosive, while satisfying the requirement that the velocity of weld propoagation, V_W, and the velocity of the flyer plate relative to the contact point, V_F, should be less than the compressive wave velocity in the two materials being welded. It is necessary to support the flyer plate with the minimum of constraint, and this can be done by providing distance pieces of wood or plastic round the periphery of the plate as shown in Fig. 6.9a or by suspending the flyer plate on cords or wire as indicated in Fig. 6.9b.

The arrangements of Fig. 6.9 are perfectly adequate for relatively small plates, but there are two major problems for larger plates. Firstly, if it is a large and relatively thin

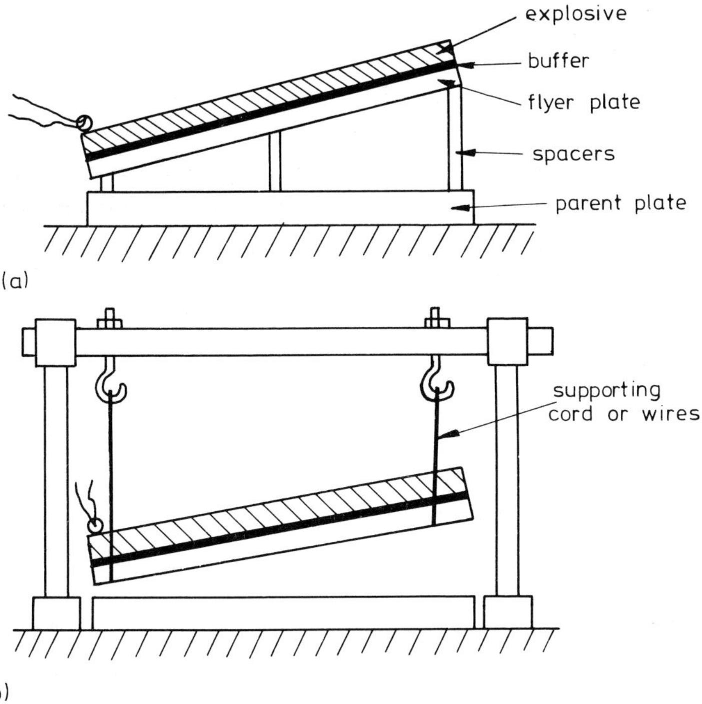

FIG. 6.9 Methods of supporting the flyer plate in the inclined set-up

plate, then there will be an excessive deflection of the plate under its own weight and the weight of the explosive, and this will cause a change of geometry from the edges to the centre of the plate, which will affect the weld quality. There is no effective way of supporting the plate away from the edges, without introducing a major obstruction between the flyer and parent plates, or attaching a wire or cord to the flyer plate, which would pass through the explosive charge and prove difficult to attach to the plate without introducing a stress discontinuity. Secondly, and more importantly, the stand-off remote from the point of initiation of detonation is large, and serious end effects result. In this region there tends to be excessive melting with separation of the weld, and the flyer plate tends to fracture. This is confirmed by Chudzic (117), who states that a stand-off distance of greater than 25 mm is undesirable due to end effects, referred to as 'flap', produced in the accelerating flyer plate.

Ruppin (118) considers that a flyer plate does not collapse in the ideal steady state fashion depicted in Fig. 6.10a, and he describes a mechanism of unstable edge movement, or flap. The conditions at the edge of the flyer plate are different to those at the centre, as the explosive is not as confined. As a consequence the detonation velocity, and the momentum imparted to the edges will be less, and the velocity of the flyer plate will be less than at the centre of the plate. This will cause the edge of the plate to curl back as envisaged in Fig. 6.10b. The net result is a 'whip lash' effect and an increased

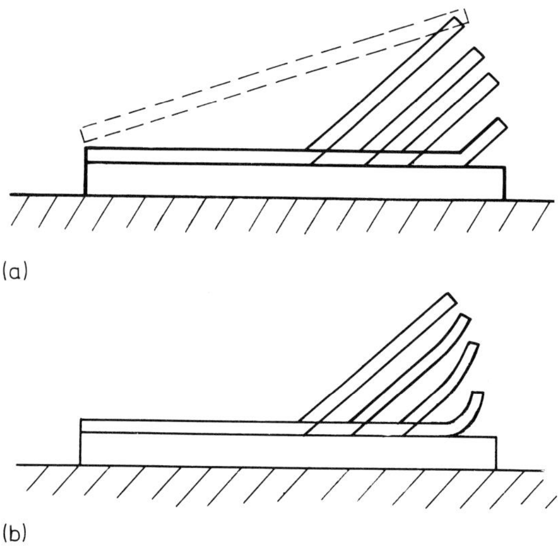

FIG. 6.10 Assumed and actual displacement of flyer plate (118)

angle of obliquity. This effect becomes more severe as the plate size increases, as for a given initial angle of obliquity the stand-off at the edge of the plate remote from the detonator will be bigger. Consequently it must be concluded that the inclined arrangement is not applicable to large plate cladding, and this is verified in practice.

To avoid edge instability it is necessary to keep the stand-off small, and this leads to the parallel method of plate cladding patented by Cowan et al (14), in which the flyer and parent plates are separated by a parallel stand-off. In the parallel arrangement the weld velocity, V_W, and the velocity of the flyer plate relative to the point impact, V_F, are equal to the detonation velocity. Consequently, if V_W and V_F are to be less than the compressive wave velocities in the parent and flyer plate materials respectively, then the detonation velocity must be less than the lower of the two compressive wave velocities. This means that a low detonation velocity explosive such as Trimonite 1 or Anfo or a mixture of explosive with a non-reactive material must be used, and such explosives require a booster charge to ensure that stable detonation is quickly achieved. With these low detonation velocity explosives the shock pressure is much less than with high detonation velocity explosives, and consequently the buffer required to prevent damage to the top surface of the flyer plate can take the form of a coat of thick plastic paint instead of a sheet of rubber or plastic. Prior to cladding both the surfaces which are to be welded together must be thoroughly cleaned; normally the surfaces are abraded with a flexible belt grinder, which removes any excessive surface contamination and gives the required surface finish.

If the flyer plate is relatively thick, so that it does not deflect significantly under its own weight and the weight of explosive, then the flyer plate can be supported at its edges by distance pieces of wood or plastic to give a parallel stand-off. Wylie and Crossland (119) used the arrangement shown in Fig. 6.11, in which the flyer plate is supported by

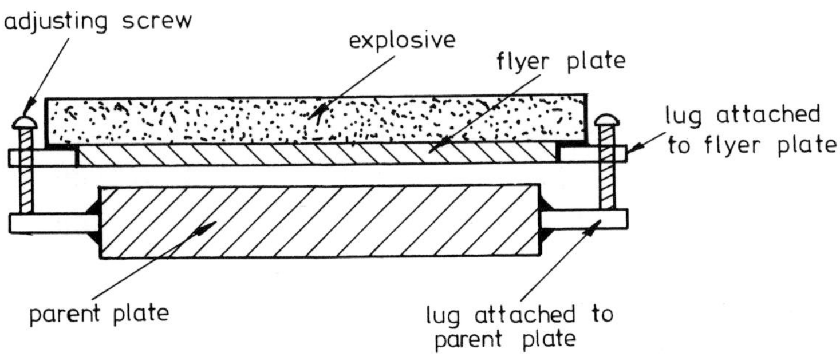

FIG. 6.11 Practical set-up for parallel arrangement for cladding (119)

screws through lugs fixed to the flyer plate which bear on lugs welded to the parent plate. The magnitude of the stand-off normally lies between the flyer plate thickness and twice the thickness, but if the minimum charge thickness to substain stable detonation gives excessive kinetic energy for a particular application, then the stand-off can be reduced so that the flyer plate does not reach its terminal velocity before impact. In order to provide a cleanly sheared-off edge, the flyer plate should normally overlap the parent plate by a distance of twice the flyer plate thickness as shown in Fig. 6.11. To reduce the edge effect the explosive layer should extend beyond the edge of the flyer plate. This may be achieved in several ways, but for a thick flyer plate a Z section steel strip can be attached to the edge of the flyer plate by screws as indicated in Fig. 6.12a, or with a thin flyer plate an angled section can be rivetted to the plate as indicated in Fig. 6.12b.

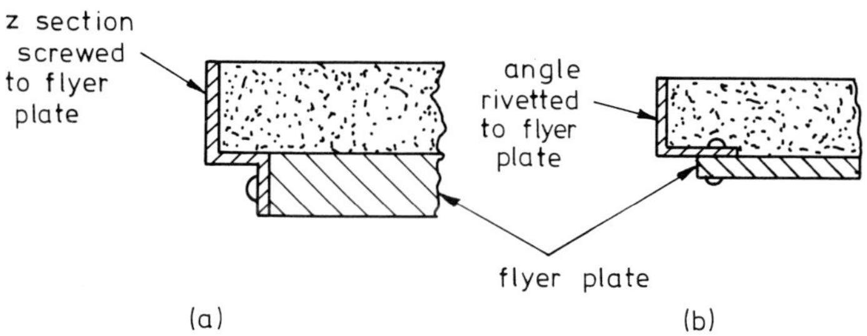

FIG. 6.12 Method of forming tray to hold explosive

The explosive used is normally a dry free running powder, and particular care should be taken that it is homogeneous so that it detonates uniformly, and consequently it should be well mixed. The explosive must be uniformly spread without any local variation in density due to the application of local pressure while spreading the explosive. This may appear to be a simple problem, but it is not so simple as it may seem to spread explosive uniformly over a large area of 10 m^2 or even greater. The explosive can be levelled with a scraper resting on the edge of the tray formed by the steel section attached to the edge of the flyer plate, but it is easy to achieve some random local compaction of the charge. This may be improved by using a scraper which is in the form of a comb.

The explosive charge may be detonated in various ways as discussed by Wylie and Crossland (119). In particular detonation may be initiated at a point, or along a straight line if a plane wave generator is used. However, there appears to be no advantage in the additional complication of a plane wave generator, so a point initiator is normally used. An adequate booster charge of some high detonation velocity explosive is required to ensure the rapid establishment of stable detonation. Frequently a cylindrical booster charge of PETN is used, or alternatively a plastic explosive or Metabel sheet explosive can be moulded into whatever shape is required.

The detonator and booster charge can be located in various positions, but it is important to realise that there is always a small area of no-bond immediately underneath the booster charge where normal impact occurs. If a square plate is being cladded, from which a circular disc is to be cut, then detonation can be initiated from a corner as indicated in Fig. 6.13a, as the corners will be removed. Frequently it does not matter if there is a small area of no-bond at the centre of a cladded circular disc, and if so the charge can be detonated from the centre. In this case a circular parent plate can be used, with either a slightly larger circular flyer plate or a square plate, as indicated in Fig. 6.13b and c. In a rectangular plate the point of initiation may be at the centre of the long side, as shown in Fig. 6.13d, which reduces the distance which the

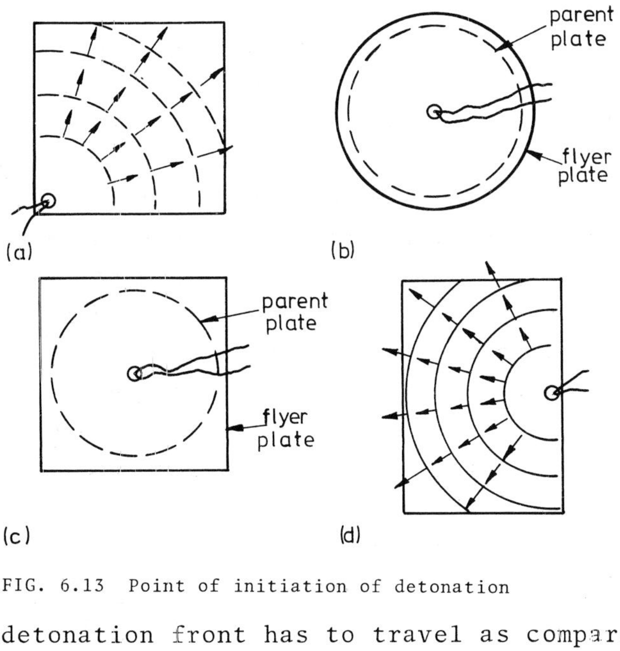

FIG. 6.13 Point of initiation of detonation

detonation front has to travel as compared with initiation at a corner. This is an advantage, as the further the detonation front has to travel the greater will be the irregularity of the detonation front. If the front is too irregular and the cladder plate is thin, then 'rucking' of the flyer plate may occur; it may be this effect, which is the ultimate limitation to the size of plate which can be cladded.

The area of no-bond under the booster charge can be reduced, according to Bergmann et al (120), by the booster illustrated in Fig. 6.14a, which causes a local deformation of the flyer plate as shown in Fig. 6.14b. This produces a local inclined impact, which considerably reduces the area of no-bond.

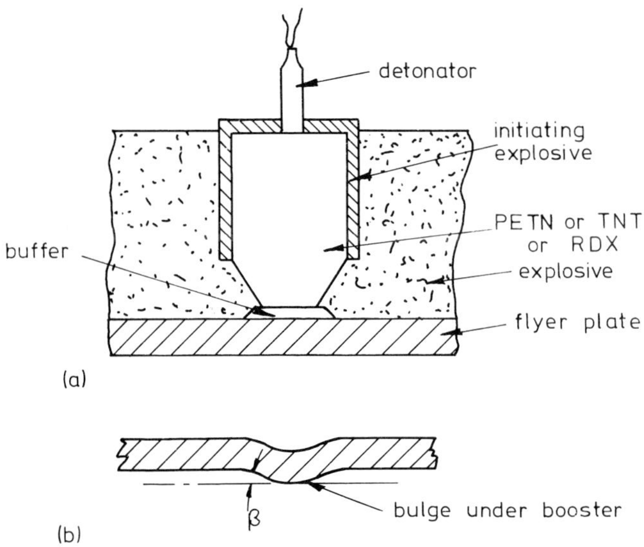

FIG. 6.14 Booster charge to reduce area of no-bond (120)

With small cladding experiments it has been usual to rest the parent plate on a thick steel anvil plate, and this is still needed if a thin parent plate is to be cladded. Though the anvil plate may require to be flattened after each operation, or after a few operations. However, in general it is not practical to have an anvil plate for most cladding operations, as it gets badly deformed or fractured. So the question is what is the best method of supporting the parent plate? The general consensus is that the best support is a thick layer of sand, and preferably moist sand. This provides a fairly uniform retarding force to the plate as the plate is forced into the sand by the detonation of the explosive charge, and consequently the plate is not excessively deformed. Nevertheless it will normally be found that the composite plate is slightly out of flatness, and if a high degree of flatness is called for then the plate must be flattened in a hydraulic press.

As has already been stated the parent plate is normally steel, and it is usually mild steel. There is always the possibility with steel, especially in cold weather on an open site, that the parent plate will fracture under the impact load imposed by the flyer plate because of a low fracture toughness or impact strength. To avoid this calamity the parent plate can be pre-heated to a modest temperature of 50°C or perhaps higher, to ensure that the temperature is well above the transition temperature. It may also help on an open site in keeping the gap between the two plates dry.

If the flyer plate is relatively thin so that its deflection when supported at the edges is comparable to the stand-off distance, then some method of supporting the plate away from the edges is required. Phillipchuk (6) suggested that the flyer plate could be supported on a knurled parent plate, as shown in Fig. 6.15, but Shribman et al (88) showed that this

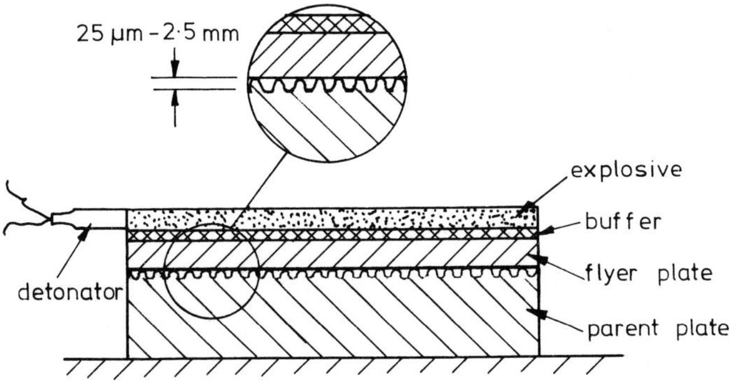

FIG. 6.15 Flat plate cladding (6)

gave a very poor weld with pockets of molten material and excessive porosity. Cowan et al (14) proposed that the stand-off could be achieved by dimples formed in the flyer plate or by tiny metal particles, as indicated in Fig. 6.16. Shribman et al (88) have shown, however, that solid metal particles are trapped at the interface where they form a barrier, which traps the jet in front of the particle and leaves a void behind. Brennecke (121) proposed to separate the flyer and parent plate by interposing thin metal ribbon deformed into a regular sine

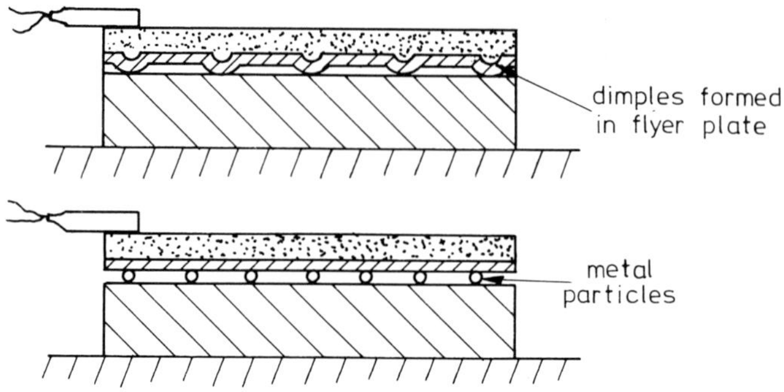

FIG. 6.16 Support of flyer plate (14)

curve configuration, which is stood on edge as shown in Fig. 6.17a or short lengths of such ribbon oriented radially from the point of initiation of detonation. Alternatively the ribbon may be twisted along its longitudinal axis. If the ribbons are arranged radially from the point of initiation of detonation, as shown in Fig. 6.17b, then it is claimed that a

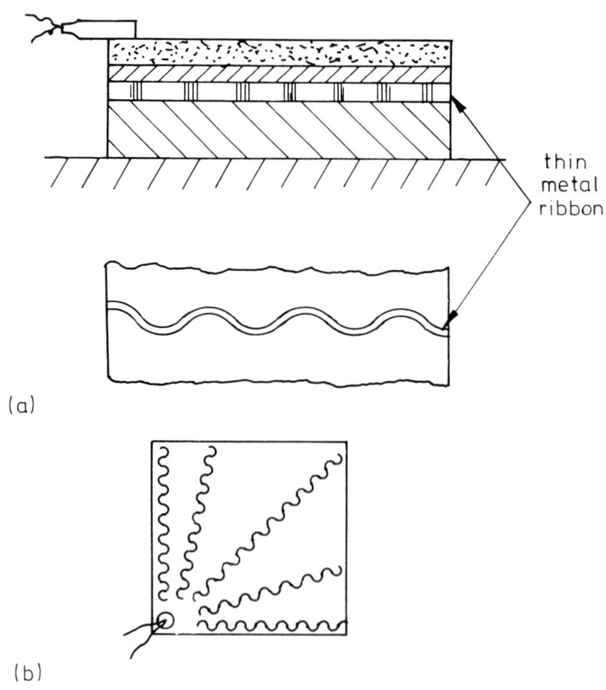

FIG. 6.17 Support of flyer plate (121)

satisfactory bond is achieved with no evidence of any defect in the bond and no blistering after heat treatment, which suggests that there is no air trapped in small areas of no-bond. Saia (122) proposed to separate the flyer plate with strips or disks of rigid foamed polystyrene or polyurethane, spaced apart across the extent of the plate. He claimed that there was no evidence of foreign matter being entrapped in the weld zone. This is confirmed by Crossland and Williams (17), who independantly used polystyrene particles to support the flyer plate. Crossland et al (123) have described yet another alternative support system, which consists of a gasket joint between the two plates arranged at the periphery with a pressurized gas supply which is adjusted to allow the plate to float on the gasket, or what might be termed a 'hover flyer plate'.

Simultaneous welding of more than two layers was proposed by Pearson and Hayes (11), who suggested in the inclined process that a third plate might be interposed between the flyer and parent plate, and which could be inclined to both. Cowan et al (14) envisaged using a parallel arrangement, with one or more plates interposed between the flyer and parent plate, and separated from one another by metal particles. There are, however, considerable limitations as to what multiple layer combinations can be welded simultaneously. Conservation of momentum dictates that after each impact there will be a charge of velocity, kinetic energy and angle of obliquity. Assuming that the kinetic energy imparted to the first flyer plate is near the upper limit of welding for the combination, then it may be that the impact velocity for one of the successive welds may be less than the lower limit for welding for that particular combination. There is also the additional complexity of assembling a multi-layer sandwich with metal particles or ribbons between each layer, so normally with multi-layer composites one layer is welded at a time.

Under well controlled cladding conditions an adequate bond will be produced up to the edge of the plate, though there may be some small amount of spalling close to the plate edge due to reflected tensile waves. If spalling is a serious problem it can be overcome by the use of spall bars which are fixed so as to be in contact with the edge of the flyer plate, and which form an extension of the plate as indicated in Fig. 6.18. The

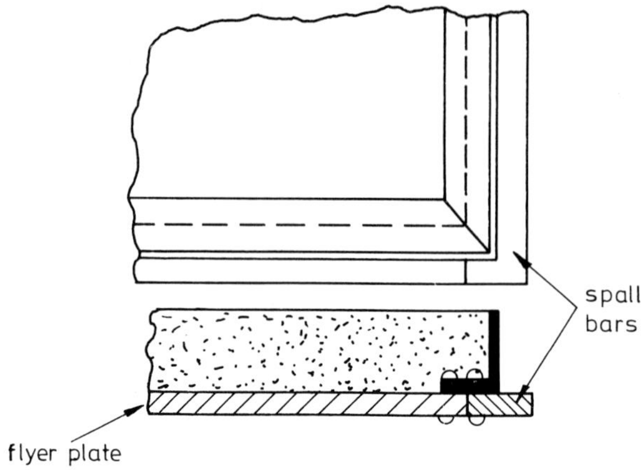

FIG. 6.18 Spall bars to prevent spalling at edge of flyer plate

compression wave enters the spall bar and it is reflected from the free edge as a tension wave, which when it reaches the spall bar/flyer plate interface causes the spall bar to separate from the flyer plate. The cladded plate will normally need to be trimmed on all edges, but 10 mm or less will usually be adequate to remove any edge defects.

Inspection of cladded plate produced by any process poses a problem, as though ultrasonic testing will establish if there is a bond it will not detect a poor bond, which might arise from an brittle cast interlayer at the weld interface. The interpretation of the ultrasonic record depends on the acoustic impedance of each of the component metals, that is the product of the density and compressive wave velocity. If the acoustic impedance of the two component metals are identical, then there will be no reflection from the weld interface if it is fully bonded. However, if there is a mismatch of acoustic impedance, then there will be a reflection from the weld interface, but if the weld interface is fully bonded then there will be a reflection from the back surface. Sometimes, as in the case of a circular plate cut from a square plate, there will be cladded material available for mechanical testing and examination. If not, and the specification demands coupons for mechanical test purposes, then a somewhat larger cladded plate must be produced.

During the cladding process it is to be expected that

residual stresses will be introduced, especially in the region of the weld interface, and the component material will become shock hardened. Consequently it may be necessary to subject the cladded plate to a stress relieving heat treatment. It is also necessary to raise the temperature of cladded plate which is to be formed in a press or reduced in thickness by rolling. It is necessary to ensure that such heat treatments do not create unacceptable metallurgical problems, such as carbon migration from a carbon steel parent plate into a stainless steel cladder plate, which would reduce the corrosion resistance of the stainless steel. Another example is the formation of brittle intermetallics at the interface of titanium/steel composite. If the diffusion across the weld interface is unacceptable it may be possible to overcome the problem by using an interlayer of a third metal, as is found desirable in diffusion bonding of some combinations. Sakhnovskaya et al (124) (125), for example, conclude that when welding industrial aluminium alloys to stainless steel it is advisable to use an interlayer of aluminium alloyed with silicon, iron, nickel or copper, which raises the permissable heat treatment temperature without the formation of harmful intermetallic compounds.

Despite the problems of diffusion and the formation of harmful intermetallics it is possible to successfully hot roll stainless steel/steel and titanium/steel cladded plates. Teslenko et al (126) have studied the heat treatment of stainless steel welded to a high carbon steel. After annealing at 650°C the composition of the melted regions was unchanged, but at 850°C there was considerable decarburization of the steel and the formation of chromium carbides along the grain boundaries in the stainless steel which reduces its corrosion resistance. These carbides can only be redissolved by an annealing heat treatment at 1200°C. Holtzman (127) has described how to overcome these problems in the hot rolling of stainless steel/carbon steel cladded plate. It is necessary for the austenitic stainless to have a carbon content of less than 0.03%, while the steel should have an average carbon content of less than 0.12% and less than 0.06% at its surface. The composite plate is then heated to above 845°C prior to hot rolling to the required thickness, it must then be cooled to less than

480°C in two hours. This is followed by descaling, then cold rolling to give a good surface finish. If the cold working is less than 2% no further heat-treatment is needed, but if it is in excess of 2% then the sheet or strip should be annealed at 900-950°C in an inert atmosphere, otherwise a final descaling treatment is necessary. It is essential to keep the rolling temperature above 845°C and to cool down to less than 480° in two hours, otherwise drastic pitting will occur during descaling unless the carbides are redissolved by a full annealing heat-treatment. De Maris (128) has described the rolling of titanium/steel clad plate and the precautions to be taken to prevent excessive diffusion. Holtzman (129) describes the rolling limitation for other metal combinations but not the heat-treatment difficulties.

Having produced cladded plates, there still remains the problem of how these can be fabricated into pressure vessels or other structures. The cladded plates can be formed into plates for both the cylindrical shell and dished ends of a vessel. However, the essential problem remains, that is how can cladded plates be welded together? The welding of stainless steel/steel cladded plate poses few problems, except to ensure that chromium carbide formation does not occur in the stainless steel weld, as this would seriously affect its corrosion resistance; Fig. 6.19a indicates the form of butt weld adopted. With titanium/steel clad plate the problem of welding is greater, as titanium and steel form a brittle intermetallic if fused together. The solution is to adopt the double-V silver inlay batten strap weld joint design described by Williams (130), and shown in Fig. 6.19b. The titanium is cut back from the edge of the plate, and the steel backing plates are then welded together. A silver inlay is then weld deposited, and a titanium strap is then fillet welded to the titanium cladding on each side of the butt weld. The problem of welding tantalum/steel cladded plates poses very considerable problems, since the melting point of tantalum is 2850°C compared with 1520°C for steel and a boiling point for steel of 3200°C. Attempts to adopt the method used for titanium clad plate, which is shown in Fig. 6.19b, failed with an acceptable thickness of tantalum (<1.5 mm), due to the eruption of liquid steel through

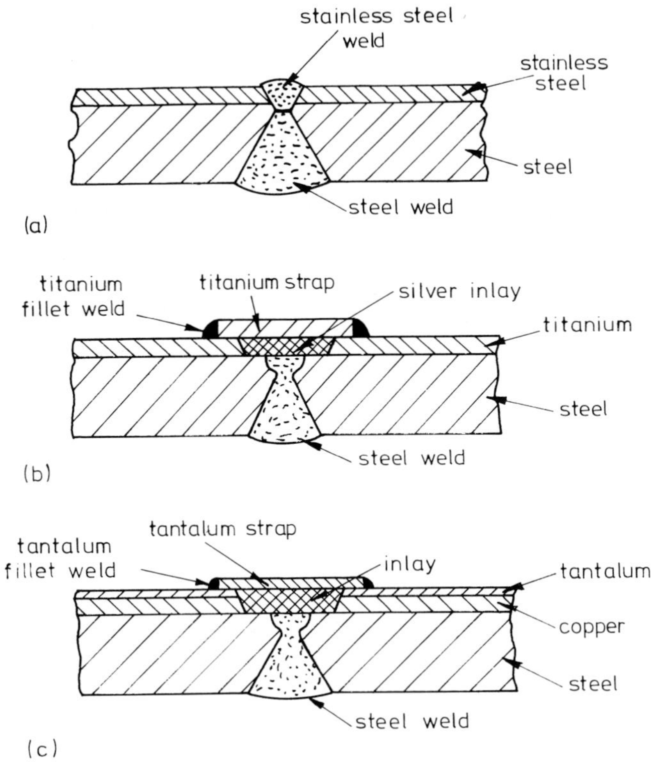

FIG. 6.19 Butt welding of cladded plate

the tantalum weld zone. If a thicker layer of tantalum is used, then a brittle structure is created at the tantalum/steel interface. The ultimate solution which is described by Bouckaert et al (131) is shown in Fig. 6.19c. A tri-clad can be produced in a single operation or by successive cladding, in sizes limited to 2.35 m by 1.04 m, though since that time plates of twice the size have been produced. The tantalum and copper are cut back, as shown in Fig. 6.19c, and the process is then similar to the butt welding of titanium/steel cladded plates. The complexities in the construction of a complete tantalum lined vessel is fully described by Bouckaert.

The problem of forming plates for fabrication into a pressure vessel, raises the question whether plates might be formed prior to cladding. Persson (116) mentions experimental work with dished ends with a radius of curvature of 1.5 m on a diameter of 1 m, in which a good bond was achieved over the complete surface. However, no indication was given of how this was achieved. Perhaps it is possible to support the shaped parent plate, and then have a flat flyer plate which is fired into the dish, so that it is deformed and welded simultaneously, but this might pose problems of satisfying the welding parameters, and avoiding flyer plate instability. Alternatively the flyer plate could be shaped to provide a parallel stand-off, when placed in the pre-formed parent plate, but this poses the problem as to how the explosive can be spread uniformly and retained. Ruppin (132) has reported on experimental investigation of the simultaneous explosive welding and forming of dished-ends. A circular disc was cladded using the parallel welding technique with initiation of detonation at the centre. The assembly was supported on a tension ring, which acted as a forming tool. If the stand-off is sufficient to allow the flyer plate to achieve its terminal velocity, so that the system can be considered as conservative, then the momentum will be conserved. This momentum must be adequate to force the cladded plate through the die. There will be limitations imposed by the need to satisfy the welding parameters, while having sufficient, but not excessive, energy to form the dished-end. Additionally there will be the problem of deep drawing without a clamping ring. Ruppin reported copper/steel and stainless steel/steel tests on blanks of 500 and 750 mm diameter.

It can be concluded that explosive cladding of flat plates is now a fully accepted commercial process, and the main limitation on the size of plate which can be cladded is that imposed by the size of charge which can be detonated without disturbing the public. Underground tunnels are very expensive unless one is fortunate to find a suitable mining complex, and even then the maintenance may be prohibitively expensive. Open sites have at present to be in remote areas with all the associate cost of transport and handling problems on an open site. However, there are developments which will reduce the blast and noise associated with the detonation of large charges, and this may allow more accessible sites to be used or allow larger charges to be detonated on existing sites. Currently the largest plates which have been cladded have areas of 30-40 m^2 and charges of up to 1500 kg or more have been used. Explosive Fabricators Inc. have simultaneously clad both sides of thick plates, which are set-up in the vertical position with open top boxes on each side to carry the explosive charges required to drive the two flyer plates.

7

WELDING OF CYLINDRICAL SURFACES

Explosive cladding of cylindrical surfaces was an early development in the field of explosive welding. Initially it was applied in particular to nozzles in steam power plant, where it is desirable to have a corrosion resistant liner. External cladding of cylindrical bars has been practised and used, for example, to produce brass cladded bar which is then machined into piston heads for hydraulic jacks and pumps. Copper/aluminium and other composite hollow cylinders have been produced by internal cladding using a massive external cylinder as an anvil, or more usually by external cladding using a core bar as an anvil in the inner tube to prevent collapse. Subsequently these cylinders are drawn down to a much smaller diameter composite tube. In the case of copper/aluminium tubing it is used for transition joints between copper and aluminium tubing in refrigerators. To form the transition joint the outer material is removed at one end of a short length of composite tubing to expose the core tube, while at the other end the inner material is removed to expose the outer tube.

At about the same time as the development of cladding of the bore and outer diameter of the cylinders, it was realised that explosive welding could be applied to joining pipes. A considerable effort has been devoted to developing the process for joining both small and very large pipes. Entirely satisfactory joints can be produced, but the problem of noise generation has not been overcome, so it is not acceptable as a process in a built-up area. However, one application where noise is not a problem is being actively pursued at the present time, that is the joining of very large pipes under water at a fairly considerable depth which has obvious application to the North Sea oil fields. Another very similar application which has been explored is the welding of collars onto pipes to provide a transition joint between a tube and a thermal sleeve.

More recently it was realised that explosive welding could be applied to the welding of tubes into tubeplates. There are many situations where a roller-expanded tube is insufficiently reliable and fusion welding is either difficult or impossible. In such situations explosive welding is now a fully accepted commercial process, which allows a large range of tube and tubeplate materials to be welded with good reproducibility and reliability.

The most recent application of welding to cylindrical surfaces has been to the explosive plugging of heat exchangers. In heat exchangers of conventional and nuclear power plant there are a large number of tubes and tube to tubeplate welds, and some failures of tubes and welds must be expected in the lifetime of the plant. The replacement of tubes or repair welds can result in a long and consequently expensive down-time, and it has been conventional practice to block-off the faulty tube with a fusion welded plug or a plug which is an interference fit. However, a mechanical interference plug is not acceptable in many situations and the access for fusion welding may be difficult or impossible, and in any case it is difficult to adequately inspect a fillet weld round a plug. Explosive plugging overcomes these problems and it is now practised in the U.K. and elsewhere.

The explosive welding of concentric cylinders was perhaps first mentioned by Phillipchuk (133) and Holtzman and Rudershausen (134) and it was patented by Popoff (135). Since that time the subject has been extensively studied, and several patents have been granted and numerous papers and patents have been published including Buckwald (136), Barrington (137), Dalrymple and Johnson (138), Doherty and Knop (139), Addison et al (140), Schetky (141), Blazynski and Dara (142) (143) (144), Crossland et al (145) and Yoblin and Mote (146).

Basically there are three systems that have been successfully employed, which are illustrated in Figs. 7.1, 7.2 and 7.3. In the first of these the explosive charge is placed inside the inner tube, which is separated from the outer tube by an annular clearance. The outer tube is supported in a massive cylinder which acts as an anvil. To prevent the external tube welding to the anvil cylinder it is necessary to use some separating medium such as a coat of paint or oil. As it is a

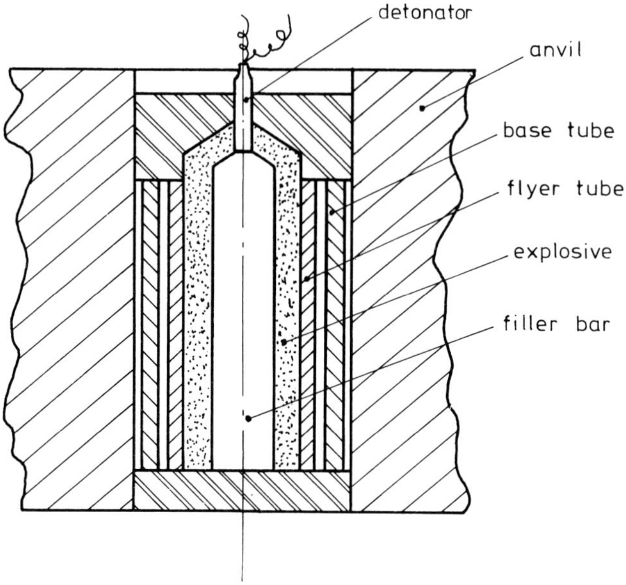

FIG. 7.1 Explosive welding of tubes

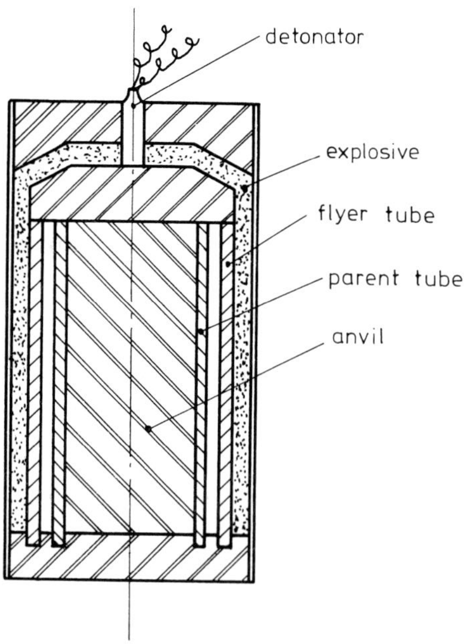

FIG. 7.2 Implosive welding of tubes

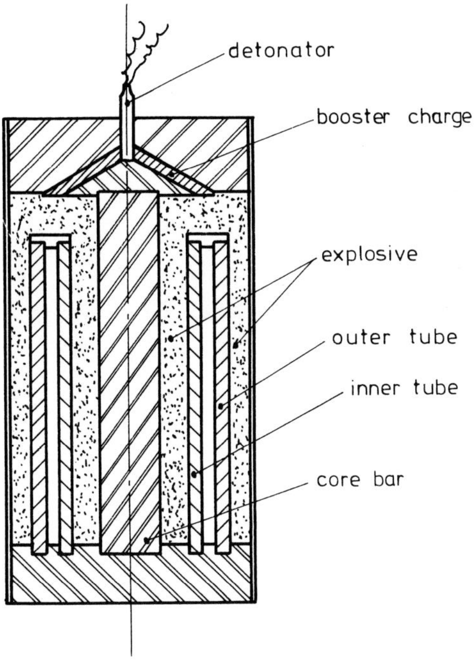

FIG. 7.3 Explosive/implosive welding of tubes

parallel explosive arrangement, it is necessary to use an explosive with a detonation velocity less than the compressive wave velocity in either the inner or outer tube material. Such explosive normally needs a booster charge of some high detonation velocity explosive in order quickly to establish stable detonation. In the case of the explosive system shown in Fig. 7.1 Buckwald (136) noted that the desired bonding conditions were not achieved for the first 50 - 200 mm of the weld length, which he attributed to the lower impulse due to lack of confinement of the charge. In order to overcome this he suggested that the explosive at the top end of the charge and extending for 50 - 200 mm should either be a different combination or more densely packed to give an enhanced impulse.

The parameters to be satisfied to achieve an adequate weld are more or less the same as those discussed in Chapter 5. However, it must be noted that the charge is more confined than for a flat charge, and consequently the detonation velocity and the velocity imparted to the tube will be somewhat higher. The latter effect may be partially off-set by the greater plastic work expended in expanding the tube, particularly if it is a relatively thick-walled tube. The explosive loading ratio, R can be varied by inserting a plastic sleeve in the inner cylinder, which also serves as a buffer between the explosive and the tube and allows the charge to be extended outside the confines of the tube. This allows a greater distance for the detonation to achieve a steady state velocity. A core tube of some inert material such as wood can be used to take up some of the internal volume. If the outer tube is sufficiently massive, as in the case of a nozzle, then an anvil cylinder will not be required. Fig. 7.4 shows some typical nozzles which have been cladded at the bore.

FIG. 7.4 Nozzle clads (Asahi Chemical Industry Co. Ltd.)

In the second arrangement shown in Fig. 7.2 the charge is in the form of a cylindrical charge surrounding the outer tube; this is referred to as an implosive system. The inner tube has to be supported to prevent it collapsing, and this may be achieved by using a core bar of a low melting point alloy, which can be melted down and re-used, or of wood. An alternative suggested by Barrington (137) is to fill the inner tube with a liquid, while Schetky (141) suggests a mixture of water and sand. With implosive welding the charge is not normally confined as with the explosive system, so it behaves in a very similar manner to a flat charge as used in flat plate cladding. For a relatively thick-walled tube there will be some reduction in the terminal velocity, due to the work expended in collapsing the tube. The explosive charge can be simply contained in a cardboard tube, which is located so as to be concentric with the outer and inner tubes. However, there is a problem in achieving concentricity for long tubes, and some form of spacer between the cardboard tube and the outer tube and between the outer and inner tubes will be required. There is also the problem in loading a powder explosive to achieve a uniform explosive charge. Schetky (141) also considers that in order to prevent local ovality of the product, it is necessary to ensure that the explosive charge should be surrounded by a uniform environment such as sand. Schetky has reported producing 1200 mm lengths of compound tubes of Admiralty brass to carbon steel and 70/30 cupro-nickel to 304 stainless steel, which had a final O.D. of 82.5 mm and I.D. of 57.5 mm. These tubes were subsequently drawn down to 22 mm O.D. tubing. Doherty and Knop (139) have described the implosive welding of a 316 stainless steel tube of 20.3 mm outer diameter and a 2.4 mm wall thickness, to an inner tube of 0.5 mm thick tantalum. They used a cast metal anvil to support the inner tube and they have successfully produced lengths of $2\frac{1}{2}$, $4\frac{1}{2}$ and $10\frac{1}{2}$ m, though for the longer lengths the tube had to be tensioned to keep them straight. Fig. 7.5 shows piston heads machined from brass cladded to steel bars, while Fig. 7.6 shows some aluminium/copper transition joints made from composite tubing drawn from an implosively cladded composite hollow.

FIG. 7.5 Piston heads machined from brass cladded steel bar (Asahi Chemical Industry Co. Ltd.)

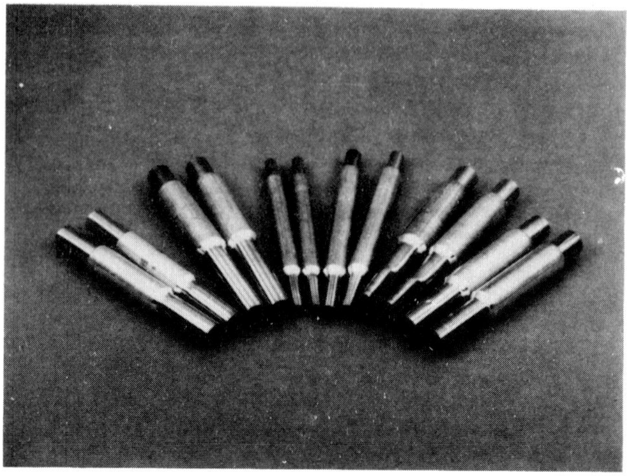

FIG. 7.6 Aluminium/copper transition joints drawn from implosively cladded tube (Asahi Chemical Industry Co. Ltd.)

In the last arrangement shown in Fig. 7.3 the need for either an internal or external anvil is avoided, by balancing the external and internal impulses generated by the inner and outer explosive charges. This might appear to be an ideal arrangement, but there are some limitations. The explosive loading ratio for the outer and inner tube should be so arranged as to impart equal and opposite impulses, but with a powder explosive this does not necessarily ensure an identical detonation velocity. So even if detonation of both charges is initiated at precisely the same instant, which is not difficult to achieve, it does not ensure that the detonation fronts in each of the charges remain in step, and this will be particularly serious the longer the tube. If the tube is too long and the inner diameter is too small, the gases produced by the detonation of the charge cannot escape quickly enough so that an excessive internal pressure can be produced, which can cause failure of the composite tube. However, if the tubes being welded are not too long, and if the bore diameter is not too small, then the process is feasible and according to Yoblin and Mote (146) it is commercially viable.

Methods of permanently joining pipes using explosives were proposed by Horeth (147), but these only involved explosive expansion of the two pipe ends into a surrounding metal sleeve to produce a tight interference fit. More recently Otto and Carpenter (148) and Capper and Chaplin (149) have described an explosive actuated method of swaging a metal sleeve on to the two pipe ends to be joined. Willis and Murdie (150) were the first to describe the explosive joining of pipes by welding the two ends to a metal sleeve. Since then a few papers have been published on the joining of pipes by explosive welding, including Verzeletti et al (151), Grollo (152), Van der Horst (153), Chadwick and Evans (154) (155) and Yoblin and Mote (146). Work has been reported by Dubin et al (156) on the welding together of pipes using a pulsed magnetic field to impart momentum instead of explosive.

Willis and Murdie (150) used the arrangement shown in Fig. 7.7, and they demonstrated the process by joining Noral

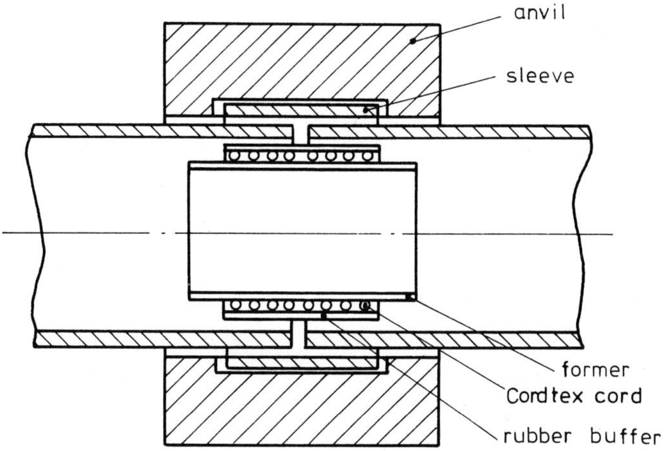

FIG. 7.7 Explosive joining of aluminium pipes (150)

50 SWP alloy pipes, which had an outer diameter of 82.3 mm and a wall thickness of 3 mm, using a Noral 50 SWP sleeve of 82.8 mm inner diameter with a wall thickness of 3 mm. The internal explosive charge consists of a cardboard cylindrical former wound with Cordtex high detonation velocity explosive cord and wrapped with rubber to act as a buffer. The charge imparts a radial velocity to the tube ends which impact the alloy sleeve. The aluminium alloy sleeve is constrained by a cylindrical steel anvil, which is made in two halves bolted together. As the detonation velocity of Cordtex is about 7 km/s, which is

greater than the bulk velocity of sound in aluminium, and it is used in conjunction with a parallel set-up with very little stand-off, it might be expected that there would be no weld. However, as will be seen when line welding is discussed in Chapter 8, the weld propagation is normal to the cord and there is no weld directly under the cord. Consequently the weld interface between the pipes and sleeve exhibits only intermittent zones of welding, and this coupled with a mechanical expansion probably accounts for the pressure-tight joints.

Howell et al (157) have used an external charge system and sleeve, and an internal expanding mandrel to provide radial support of the two tube ends. Fig. 7.8 shows the

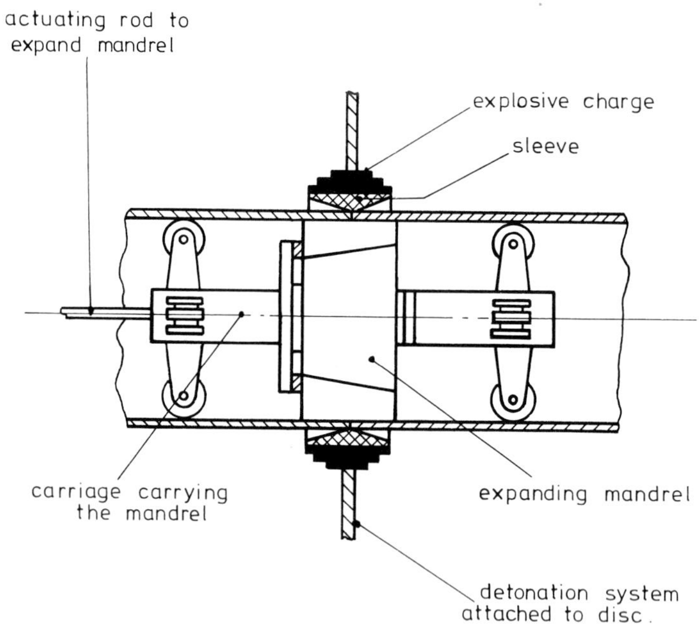

FIG. 7.8 External charge system for joining pipes (157)

arrangement employed with a double inclined sleeve which requires initiation of the charge in the central plane. Initiation of the charge must occur simultaneously round the periphery of the charge. This can be achieved, for example, by connecting the ends of equal lengths of detonating cord to various points round the periphery of the charge and connecting the other ends of the cords to a single detonator. A better system for simultaneously initiation of the charge round the periphery is described by Howell et al. Steel tubes up to 125 mm have been joined by this method though the patent claims that it can be used up to diameters of 1200 mm. Howell et al have also used a parallel sleeve as illustrated in Fig. 7.9,

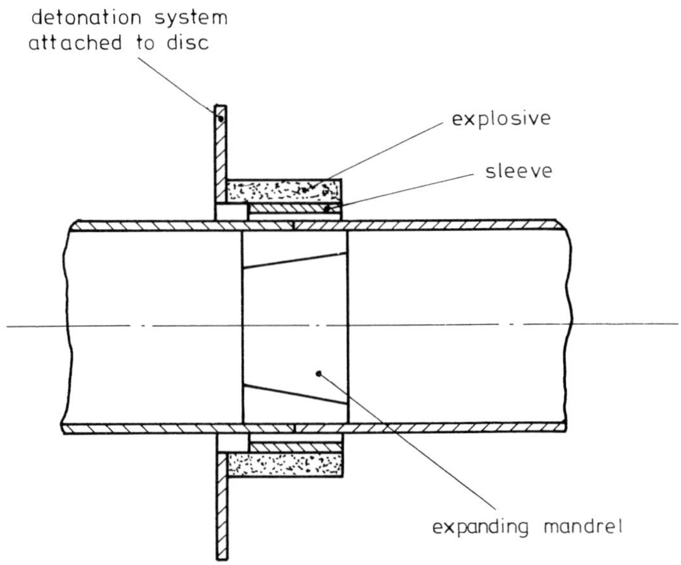

FIG. 7.9 External charge system for joining two tubes using a parallel sleeve

which requires a low detonation velocity in a cylindrical container. An axial propogating detonation wave is initiated from one end of the charge using a similar system to that described above. Using this system Howell et al has joined steel tubes having a diameter of 760 mm. There are, however, two essential problems. Firstly an expanding mandrel is required to support the tube ends, which might consist of sectors of a hollow cylinder with a tapered bore expanded radially by a tapered bolt. However, the mandrel has to withstand the severe impact imparted during the welding process. If the mandrel is too hard with a corresponding low impact strength then the sectors of the mandrel will fracture, but if the mandrel is not sufficiently hard then it will deform under impact loading and it will require frequent re-machining. The economics of the process requires the mandrel to have an adequate life. Secondly the noise generated by the charge would be unacceptable in a built-up area. For small pipes it might be possible to devise an enclosure to place round the charge while it is being detonated, though this will add to the cost of the operation, but for large pipes there is no possibility of providing an enclosure for the size of charge involved. Consequently for large pipes the external charge system is probably unacceptable except in uninhabited regions.

Chadwick and Evans (154) (155) have described the explosive welding of 27 and 89 mm nominal diameter pipes using an internal charge configuration. The arrangements used are shown in Fig. 7.10 in which the pipe ends are welded to a sleeve, which is sufficiently massive to act as its own anvil. This requires the sleeve to be seven times as thick as the pipe, in order to keep the distortion to an acceptable level. The stand-off is provided by an annular recess in the sleeve, which has a depth about equal to the thickness of the pipe. Fig. 7.10a shows the arrangement used for either a Trimonite 1 or Metabel explosive charge for welding 27 mm diameter pipes, while Fig. 7.10b shows a parallel arrangement which is only suitable for Trimonite 1. The arrangement for the 89 mm pipe is shown in Fig. 7.10c, in which axial detonation of the explosive charge of Trimonite 1 is ensured by an initiating charge of Metabel sheet explosive in the form of a disc. Initially a polyethylene container was used to hold the explosive charge, but it was found that the container was extruded between the abutting ends of the pipes into the bore of the sleeve, which prevented welding. To overcome this the explosive was contained in an outer sheath of stainless steel or mild-steel, which had a thickness of at least 0.62 mm. The sleeve has the added advantage of automatically aligning the ends of the two pipes. A much thinner sleeve can be used to join two tubes, but in this case an external segmental anvil is required to support the sleeve. Chadwick and Evans (154) used a six-segment anvil, which was restrained by a rubber lined metal container. They noted that there was some distortion of the segments, which subsequently prevented them being in intimate contact with each other or with the joining sleeve, however, this was found not to be important. They did not establish the useful life of a segmental anvil, but they con-

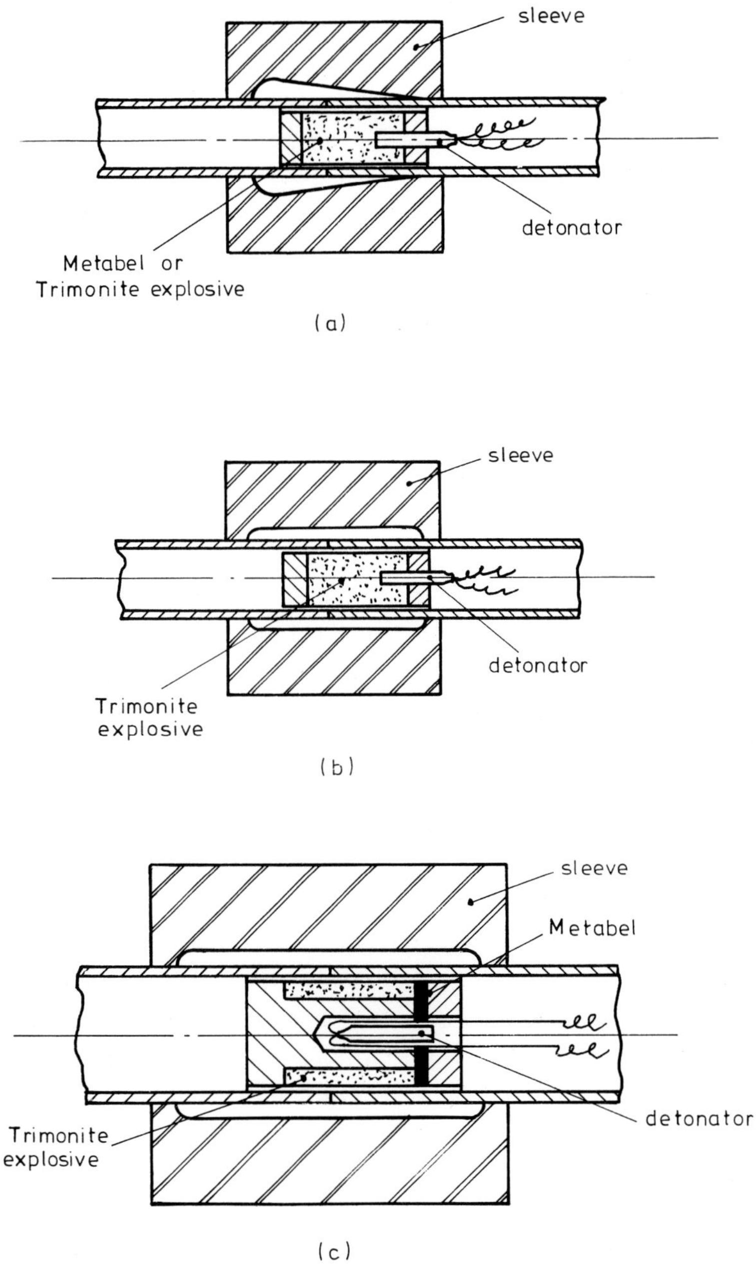

FIG. 7.10 Internal charge system for joining two tubes (154, 155)

sidered it to be more economical than using thick-walled sleeves.

Chadwick and Evans have also reported the welding of

pipes of 203 mm diameter by 9.5 mm wall thickness using the arrangement of Fig. 7.11. This arrangement was adopted be-

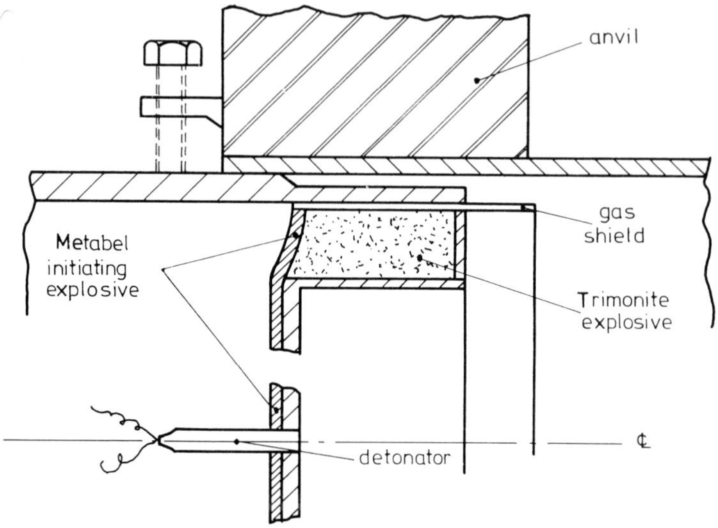

FIG. 7.11 Configuration used for 203 mm tubes (154)

cause those shown in Fig. 7.10 require a sleeve which has a length two or three times the diameter of the pipe to avoid bulging the pipes outside the confines of the sleeve, and this would require an expensive and unwieldy sleeve for a large diameter pipe. The arrangement requires that the wall of the pipe must be thinned down from 9.5 to 5 mm to provide an adequate stand-off. The mild steel support anvil was 355 mm diameter by 101 mm long, the bore of which increased by 1.5 mm after the first welding operation. A six segmental anvil with a similar outer diameter was also tried, and dents of 1.3 mm deep corresponding to the weld zone were noted. No significant bulging of the pipes outside the confines of the anvil was noted.

More recently International Research and Development Co. Ltd. have extended the work reported by Chadwick and Evans to welding together pipes of 762 mm by 12.7 mm wall thickness, and 812 mm diameter by 19.1 mm wall thickness. The set-up which is shown in Fig. 7.12 is basically that shown in Fig. 7.11 but with a massive segmental anvil. Chadwick (158) has reported successful tests on welding similar sizes of pipes underwater at depths of 120 m, which was the maximum available at the test-

FIG. 7.12 IRD/Vickers system for underwater tube to tube welding

ing site. The work is aimed at producing a method of repairing underwater pipe lines.

The advantages claimed for using an internal charge arrangement over the external charge set-up are that noise is reduced, the blast debris is contained, and external support appears to be more attractive than an internal mandrel. Not only is the noise of an internal charge reduced but the noise is more amenable to damping methods, though these must not cause excessive pressure in the pipe. One of the problems of an internal charge system for large pipes is how to keep the charge in contact with the bore of the flyer tube especially if it suffers from out-of-roundness. This has been overcome according to Chadwick (158) by using an inflatable charge.

Yoblin and Mote (146) have overcome the need for an internal mandrel or an external anvil by using an external sleeve and explosive charge with a reaction charge inside the pipe. If this can be done reliably and repetitively then it overcomes a major problem, but the noise problem from the external charge still remains.

A similar system to that developed by Chadwick and Evans was developed independently by Wylie and Crossland (159) to weld a collar to a tube. The objective was to provide a means of welding a tube in a nuclear reactor to a thermal sleeve of a different material, by explosively welding a collar of the same material as the thermal sleeve to the tube, and then machining an appropriate fusion weld preparation in the collar. It was essential to achieve the welding of the collar to the tube with no crevice at either end. The arrangement finally developed is shown in Fig. 7.13 and a sectioned example of the

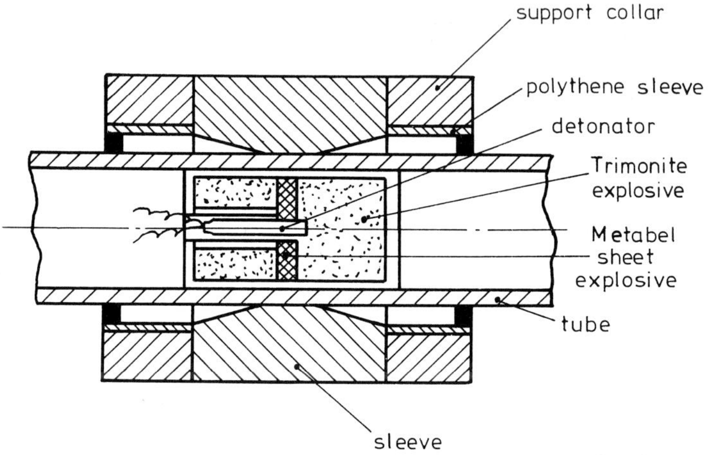

FIG. 7.13 Welding a collar to a tube (159)

product is shown in Fig. 7.14a. The problem was that if the weld was to extend up to the ends of the collar, then the tube would have sufficient energy to shear itself off against the collar. This was overcome by fitting two polythene sleeved support collars. As can be seen from Fig. 7.14b the weld continues up to the end of the collar with no crevice, and there is no shearing of the tube.

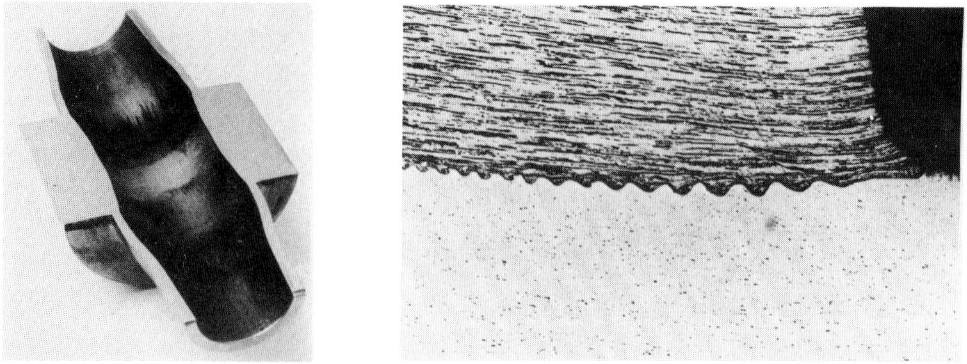

FIG. 7.14 Collar explosively welded to tube (159)

Explosive expansion of tubes into tubeplates has been proposed by Horeth (147), Mottram (160), Berman et al (161) (162) and Brown and Dunmore (164); these are illustrated in Fig. 7.15.

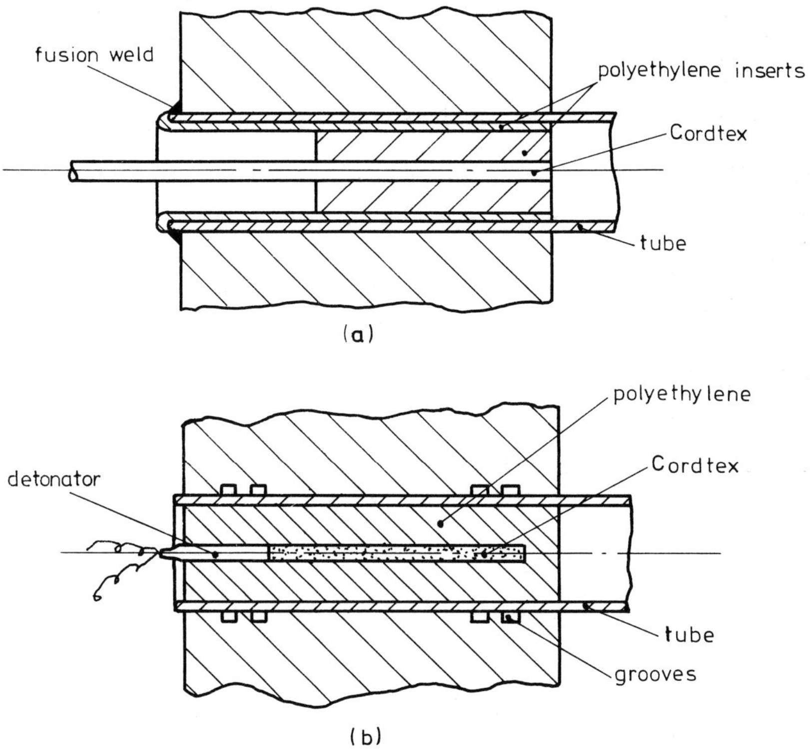

FIG. 7.15 Explosive expansion of tubes into tubeplates (160, 161)

The possibility of explosive tube to tubeplate welding might be considered as having been envisaged by Holtzman and Rudershausen (134), nevertheless it does not appear to have been explicitly mentioned until a paper by Crossland et al (165). At the same time Yorkshire Imperial Metals Ltd. (166) were developing the inclined process which has been exploited under the trade name of YIMpact.

Crossland et al (61) (165) reviewed the various arrangements for welding tubes to tubeplates and they concluded that there were advantages in using a low detonation velocity explosive, and as a result they developed the parallel arrangement shown in Fig. 7.16. This is the basic set-up which has been patented by the Associated Electrical Industries Ltd.

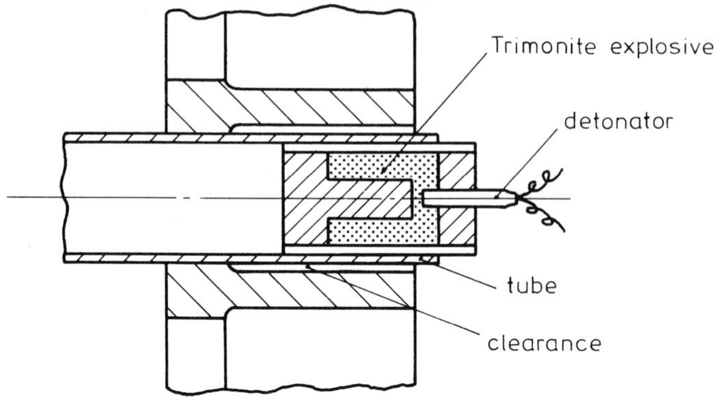

FIG. 7.16 Parallel set-up for tube to tubeplate welding (165)

(167). However, this arrangement requires the manufacture of special cartridges, and the tube when welded to the tubeplate has a step in the bore which may be undesirable. These disadvantages are overcome by the inclined process described by Chadwick et al (50), Cairns et al (168) (169) and Hardwick (170) (171) (172) which is shown in Fig. 7.17. This process

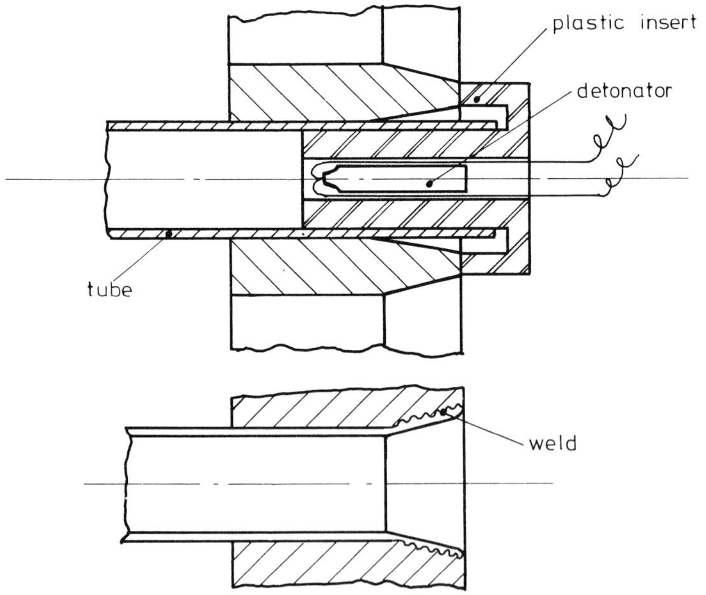

FIG. 7.17 Inclined set-up for tube to tubeplate welding (YIMpact)

uses a high detonation velocity explosive, and for small tubes a detonator alone is adequate, but for larger tubes the base charge in the detonator can be supplemented by cylindrical charges of pentalite or by a plastic sheet explosive moulded into a cylindrical charge. When the charge is detonated the length of tube standing proud of the tubeplate is sheared-off and the resultant product forms a bell-mouthed entry into the tube, which is ideal from flow considerations. Fig. 7.18 shows an example of YIMpact welding applied to a heat exchanger.

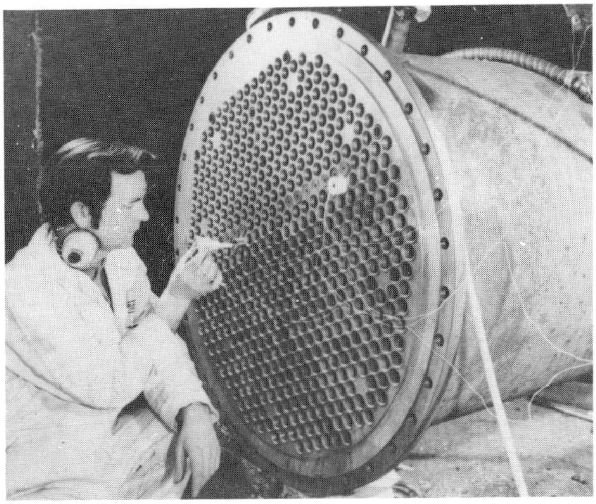

FIG. 7.18 Monel metal tubes being welded to a monel/steel explosivelycladded plate (Yorkshire Imperial Metal Ltd.)

In the YIMpact welding process the initial angle of the tapered hole is normally in the range of 10 to 20°, though the dynamic angle will be appreciably greater. The larger initial angle requires a tube material which can withstand the associated strain, and it also leads to a reduction of the ligament thickness between adjoining holes which may be unacceptable. Consequently an angle of 10° is normally employed. Surface finish and cleanliness is important; a roughness of 5 μm may lead to excessive interfacial melting which might be unacceptable for certain metal combinations. However, a surface finish of 1 μm or better is easily and cheaply achieved and this appears to be generally acceptable in practice. Though the process is reasonably tolerant of contamination, nevertheless the greater the contamination the larger is the explosive charge which is required, and this could lead to excessive ligament distortion. Consequently it is desirable that the oxide layers and grease should be removed shortly before welding, but only to a standard which can be readily and cheaply achieved within a normal workshop.

Another method of welding a tube to a tubeplate has been used by Bahrani et al (178) for welding plugs into heat exchangers, which consists of a point or small spherical charge as illustrated in Fig. 7.19. When the charge is detonated it

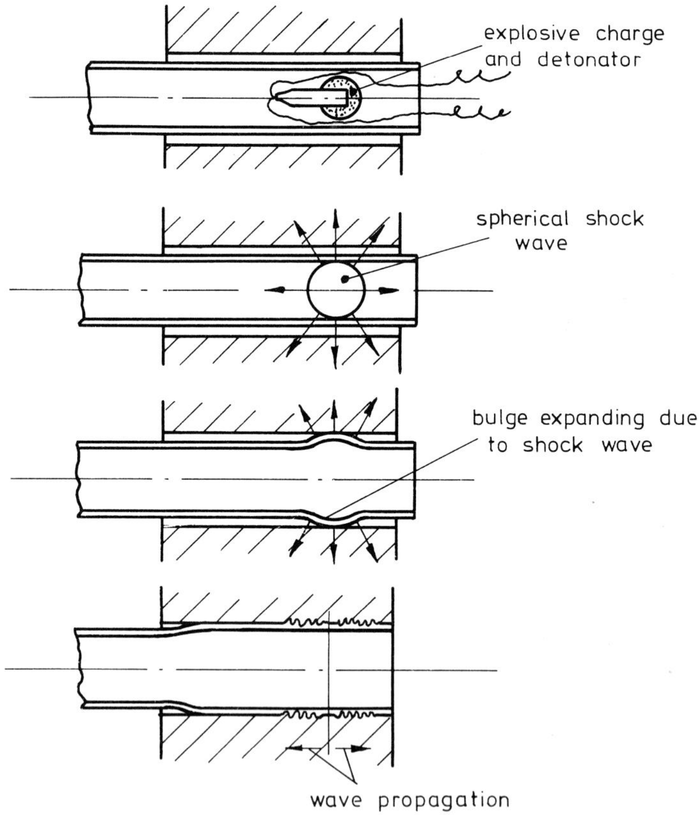

FIG. 7.19 Point charge set-up for tube to tubeplate welding (178)

produces a roughly spherical bulge in the tube wall. At the centre-line of the bulge the impact is normal to the bore of the hole, so no welding will take place. However, each side of the bulge there is a high velocity oblique impact which produces two zones of welding. If the centre of the charge is correctly positioned the excess length of tube projecting beyond the front surface will be sheared-off and a weld can be produced up to the front surface of the tubeplate. For bigger bores the bulge produced by a point charge does not give a sufficiently large angle of obliquity close to the centre-line of the bulge, and to overcome this Crossland et al (183) have used a flywheel shaped charge of high detonation velocity explosive as illustrated in Fig. 7.20. This produces a bulge in the tube wall which in the longitudinal plane has a much smaller radius than for a point charge, and consequently adequate angles of obliquity for welding are generated closer to the centre-line of the bulge. The flywheel charge has the additional possibility of achieving a satisfactory weld with a thinner tubeplate, without causing bulging of the tube behind the tubeplate.

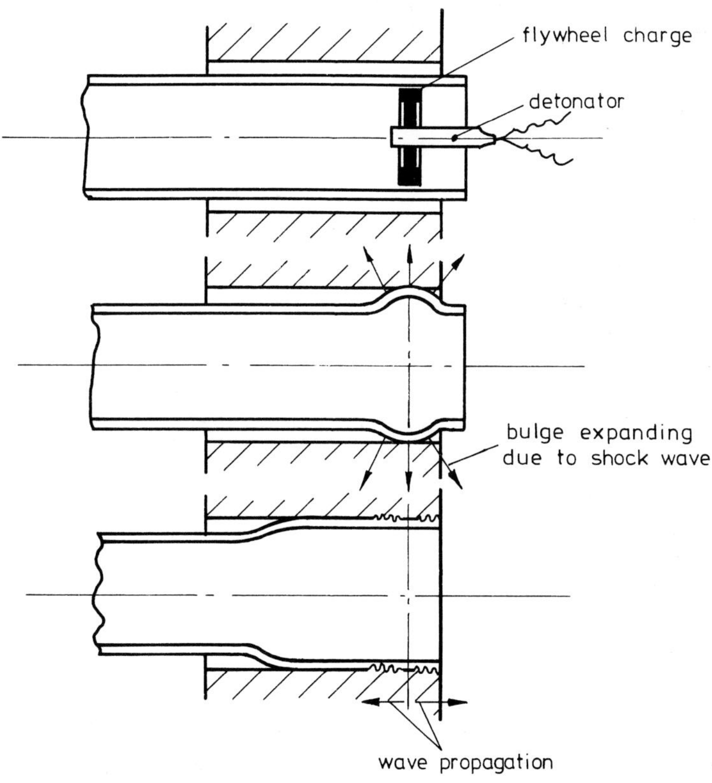

FIG. 7.20 Flywheel charge set-up for tube to tubeplate welding (179)

Simultaneous welding of a number of tubes is possible, but if adjoining tubes are to be welded then it is desirable to have simultaneous detonation of both charges, which cannot be achieved with standard detonators. It can be achieved as shown by Shribman et al (173) by connecting each charge by equal lengths of detonating cord to a common initiation source. Even then there will be an enhanced tensile stress generated in the ligament at the point where the shock waves from adjoining tubes meet. These problems could be overcome by welding alternate tubes at the same instant, but the complexity of the wiring and the greatly enhanced noise level are both major disadvantages. Normally one tube end at a time is welded and with good organisation a high repetition rate can be achieved, and the noise level is such that it can be simply accommodated in a factory environment.

If the ligament is excessively small then unacceptable distortion will be experienced in holes adjoining the one in which a tube is welded. This may make it difficult or impossible to satisfactorially weld tubes in the adjoining holes. Hardwick (172) has noted this limitation which he regards as being a function of the ratio of ligament thickness to tube wall thickness. If this ratio is too small, then the distortion will be unacceptable, and it will then be necessary to locally thin down the tube wall in the region of the weld zone to reduce the distortion. Shribman et al (173) and Crossland and Williams (174) have studied the distortion of the adjoining holes, and from a dimensional analysis of the parallel tube to tubeplate welding process they deduced that the radial distortion of the adjoining holes is given by:

$$\Delta R = \frac{K^1 \rho V_t^{\,2} D^2 t^{\frac{1}{2}}}{L^{\frac{3}{2}}}$$

where,

ΔR is the radial distortion in mm

$K^1 = \frac{K}{\sigma}$, where K is a constant

σ is the yield stress of the tubeplate material in $\frac{MN}{m^2}$

ρ is the density of the tube material $\frac{kg}{m^3}$

V_t is the radial velocity imparted to the tube in m/s

D is the diameter of the tube in mm

t is the wall thickness of the tube in mm

L is the ligament thickness in mm

For steel tubes welded to steel tubeplates this equation gives excellent correlation with extensive data available using a value of K of 3.9×10^{-9}. However, using the same value of K as determined from the steel data, the equation did not give such good correlation with the limited experimental data for welding steel tubes to aluminium and copper tubeplates. The correlation was much improved if the density of the tubeplate material was taken into account in the dimensional analysis. Crossland and Williams suggested a strategy for determining the minimum ligament thickness which could be welded.

The problem of ligament distortion can be largely overcome by using steel support bungs in adjoining holes, which for parallel holes should be slightly tapered and a small clearance fit in the hole. These bungs reduce the distortion to an acceptably small level and they can be freely removed. However, it is also necessary to place support bungs in holes in which tubes have already been welded, but as this bore is not so well defined it may mean that special plugs must be made, which might normally be commercially unacceptable.

The remaining question is what limitation is there on the thickness of the tubeplate? If the tubeplate is too thin in relation to the tube diameter, then there will be excessive bulging of the tube behind the tubeplate due to the shock wave pressure when the charge is detonated. This bulging, even if it does not result in the failure of the tube, will prevent a forward extraction of the tube if this should ever be necessary. There appears to have been no systematic study of this, though Otto et al (175) state that tube diameter to tubeplate thickness ratios commonly used in explosive welding are in the range 0.5 to 1.5. It is probable that with the arrangements so far described, with the possible exception of the point and flywheel charge, there would be considerable difficulty in avoiding unacceptable bulging with a ratio as high as 1.5, while in many applications the ratio is less than 0.5. Otto et al described a few feasibility studies on welding a 63.5 mm outer diameter by 1.65 mm wall thickness aluminium tube to a 25.4 mm thick aluminium tube sheet, which gives a ratio of 2.5. They used detonating cord wrapped round a wood plug as shown in Fig. 7.21, but much more work is required to establish

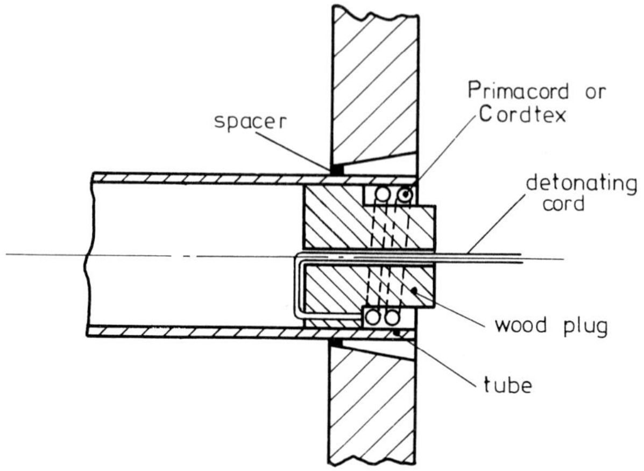

FIG. 7.21 Welding system for tube welding to thin tubeplate (175)

the applicability of this arrangement to this and other combinations. Prümmer and Jähn (176) have used an inertial plug behind the explosive charge to reduce the pressure pulse experience by the tube, as shown in Fig. 7.22.

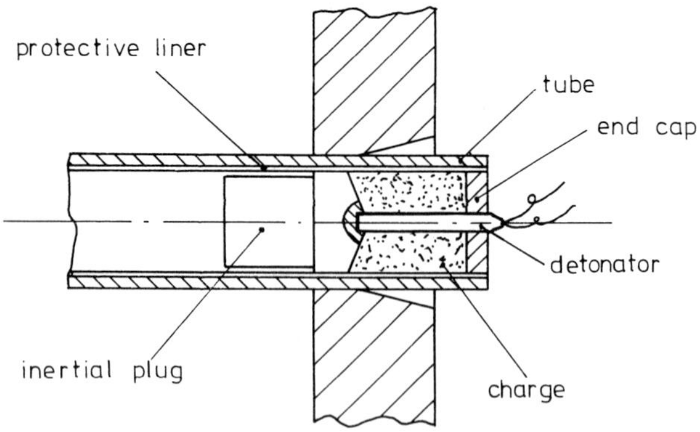

FIG. 7.22 A proposed method of reducing bulging behind the tubeplate (176)

One of the considerable advantages of the various tube to tubeplate welding processes discussed here, which applies equally to explosive plugging, is that non-destructive testing is possible. A sufficient length of weld is produced to allow ultrasonic inspection. Though automated inspection systems have been developed it appears to be adequate to use a miniature hand held probe to scan round the circumference and axially to establish the extent of the weld zone.

Explosive plugging has been developed simultaneously by several groups: Babcock and Wilcox Ltd. of America, International Research and Development Ltd., Yorkshire Imperial Metals Ltd., Atomic Energy of Canada Ltd. and the Queen's University of Belfast on behalf of Babcock Power Ltd. Johnson (177) in 1971 was the first to publish a paper on the subject and since then there have been papers by Bahrani et al (178) (180), Dalrymple (181), Hardwick (170) (182), Crossland et al (179) (183).

Johnson (177) was concerned with plugging a nickel alloy tube of 15.9 mm outer diameter by 0.89 mm wall thickness in a nuclear heat exchanger. Fig. 7.23 shows the parallel clearance

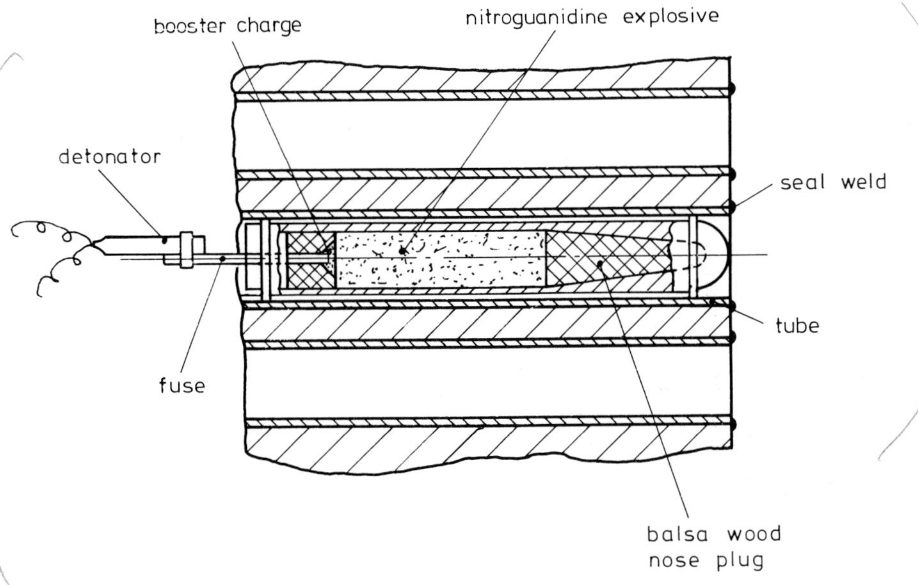

FIG. 7.23 Explosive plug (177)

plug developed for this application positioned in a tube tubeplate simulation. The plug is made from a nickel alloy and as can be seen there is a parallel stand-off. Consequently a low detonation velocity explosive is required and nitroguanidine was used with a packing density of 0.7 g/cc which gives a detonation velocity of 1.3 km/s. A satisfactory weld was created between the plug and the tube, and perhaps surprisingly there was a weld between the tube and tubeplate, even though the clearance was small and irregular. This is confirmed by Jackson and Brown (185) who claim that acceptable welds can be obtained between a tube and tubeplate with clearances as small as 0.125 mm. Johnson also describes the manually operated handling gear for the remote positioning of the plug in the tubesheet.

Hardwick (170) (182) has used a high detonation velocity explosive charge, frequently in the form of a detonator alone, as with the YIMpact tube to tubeplate welding process. However, it is difficult and costly to machine a taper in the holes to be plugged *in situ*, so the angle has to be achieved by either machining or swaging the tubular end of the plug as shown in Fig. 7.24. The machining of the plug end results in a

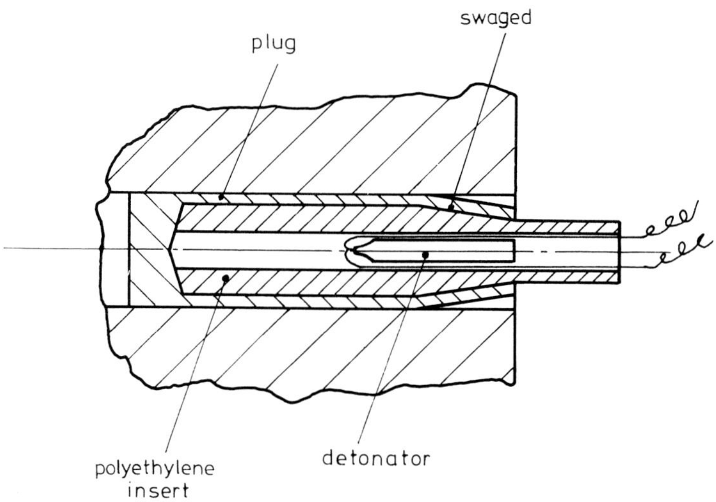

FIG. 7.24 Swaged plug (182)

tube wall thickness which varies over the length of the taper, and unless the charge is also tapered this would result in a reduction of the dynamic angle of obliquity. However, with a swaged plug the mass per unit length remains virtually unchanged even though the diameter is reduced, so a uniform explosive charge can be used. The end of the plug is swaged with a shaped polythene insert in the bore, which is secured in position by the swaging operations. The explosive charge can subsequently be inserted into the polythene sleeve. It is necessary to use a plug material which will have sufficient ductility to accommodate the expansion of the plug, and it must also be suitable for the heat exchanger application, for example in conventional power plant monel metal would be a suitable choice of material. Prior to plugging the hole is cleaned with an air operated grinder.

Bahrani et al (178) were concerned with plugging the stainless steel reheater and superheater heat exchangers of the U.K. prototype fast reactor, which were both of the toroidal design shown in Fig. 7.25. Hand access into the

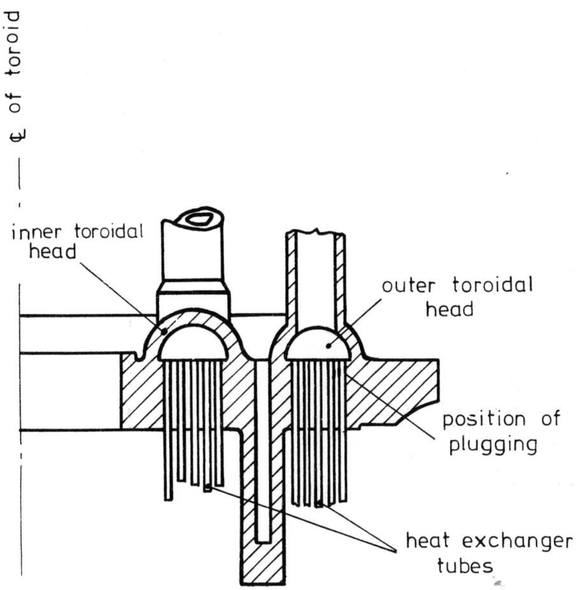

FIG. 7.25 Reheater heat exchanger of prototype fast reactor (178)

toroid was possible by cutting-off the header down pipes, and in order to be able to view the necessary manual operations a closed circuit television was employed. The head room above the holes adjoining the wall of the toroidal head is restricted to 48 mm and the tube hole diameters of the two heat exchangers are 15.5 and 20.3 mm respectively. The plug developed for this application is shown in Fig. 7.26 and it is based on the point or spherical charge system illustrated in Fig. 7.19. It will be seen that the stand-off is achieved by having a plug with a bulbous end, and locating the other end concentrically with the bore of the hole to be plugged by means of a centralising ring,

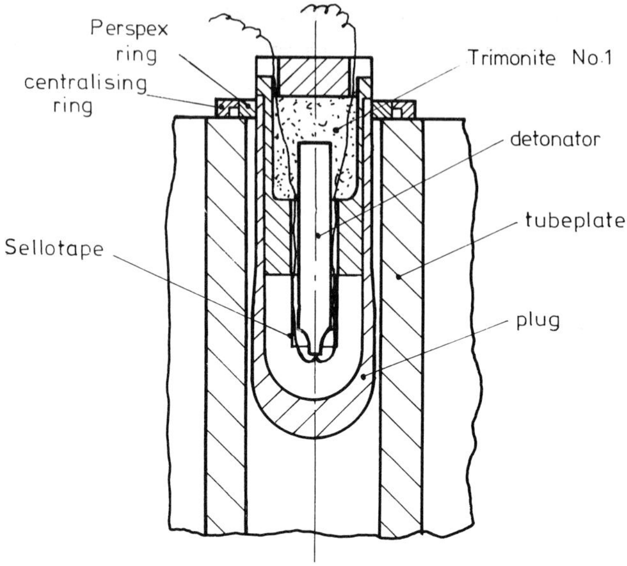

FIG. 7.26 Explosive plug for heat exchanger of prototype fast reactor (178)

which is stuck to the top surface of the tubeplate with an adhesive or double sided adhesive tape. The explosive charge consists of the base charge in the detonator supplemented by a small charge of Trimonite No. 1 explosive powder. With such a short plug there is the danger that the impulse imparted to the base of the plug could cause the plug to fail in tension, and this can be avoided by the inertia of the base of the plug. This requires that the base of the plug must be adequately thick to provide the required inertia.

In the plugging of the prototype fast reactor heat exchangers, great care is demanded as any mistake leading to a faulty plug would be extremely difficult and costly to rectify. The authors describe the great care taken in preparing the hole and sticking on the centralising ring on the top surface of the tubeplate concentric with the hole to be plugged. The plug is made of vacuum remelt stainless steel and its quality is checked by metallurgical examination of material each side of the plug. The detonator employed is of aerospace quality, which implies that it is more carefully controlled and inspected during manufacture, and the explosive is carefully sifted and kept dry prior to its use. If the schedule laid down is carefully adhered to then a high quality weld is always produced, and there is no crevice at the front surface of the tubeplate, which is an important requirement because of the potential danger of crevice corrosion with stainless steel. Operational experience on site with this plugging system has been reported by Jackson (184).

Crossland et al (179) have also been involved with the plugging of the various tubeplates of the advanced gas cooled reactors at Heysham and Hartlepool, and in particular the reheater inlet and outlet tubeplates. Fig. 7.27 shows the header

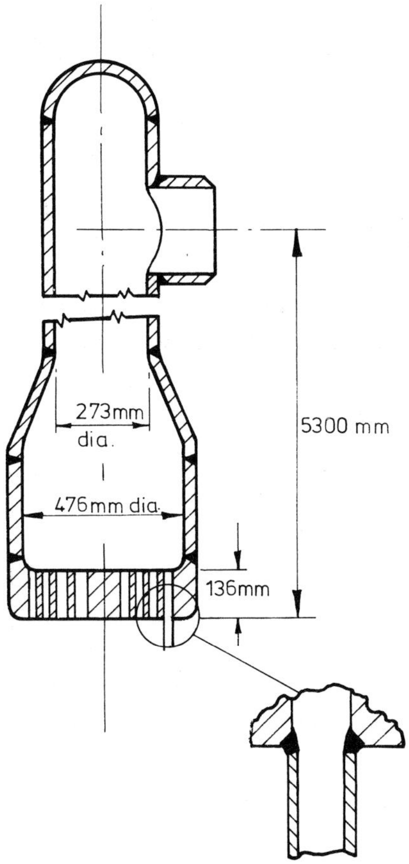

FIG. 7.27 Tubeplate and header of reheater outlet of the advanced gas cooled reactor (179)

and tubeplate of the reheater outlet which are fabricated from stainless steel. The hole diameter in the reheater inlet is 43.6 mm with a triangular pitch and a ligament thickness within the range 7.75 - 6.68 mm, and the outlet tubeplate has holes

40.4 mm diameter with a square pitch and a ligament thickness within the range 7.77 - 6.71 mm. The plug developed for this application is shown in Fig. 7.28 and it will be seen that the flywheel shaped charged illustrated in Fig. 7.20 has been used, otherwise it is similar in construction to the smaller plug used in the superheater and reheater heat exchangers of the prototype fast reactor which has already been described. However, the ligament thickness in relation to the hole diameter is small and close fitting tapered support bungs are necessary in the adjoining holes as shown in Fig. 7.28.

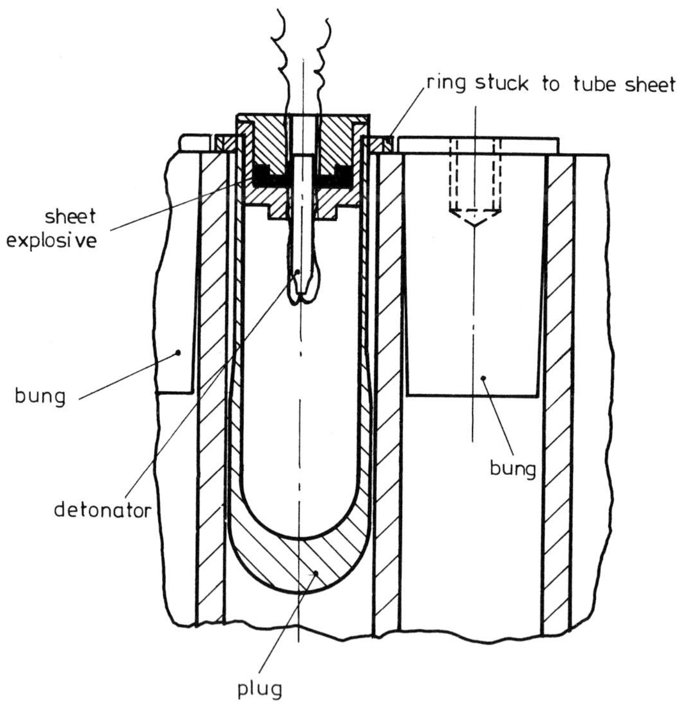

FIG. 7.28 Plug for reheater tubeplate of the advanced gas cooled reactor (179)

The major problem in this particular application was the problems of access as demonstrated by Fig. 7.27, where it will be noted that the tubeplate is 5.3 m down a 273 mm diameter header. Crossland et al (183) describe a remote manipulator which will carry out all the necessary functions controlled from a console with closed circuit television. This manipulator with its attachments can locate which tube is leaking and the associated hole in the tubeplate is then honed to remove excess contamination and to give the required surface finish. Following honing the hole is cleaned and its diameter measured to make sure that the bulbous end of the plug will fit in it. Then a centralising ring is stuck to the top surface of the tubeplate concentric with the hole to be plugged. Support bungs are placed in the holes adjoining the one to be plugged, and protection bungs to prevent the ingress of debris are placed in all the other holes. The plug with its explosive charge is then placed in position, and the manipulator is then lifted up the header trailing the detonator leads behind it, which remain shorted until they are connected to the exploder. For additional safety the aerospace detonators used have high energy fuse heads. Following the detonation the manipulator is lowered down the header with a vacuum cleaner attachment, which removes all the debris except some large pieces which are picked up with a grab. All the bungs are then removed and a leak detection head is used to check the welded plug. The welding process requires a 0.75 mm 45° chamfer on the lip of the hole to be plugged to eliminate a crevice, and it may also be necessary after welding to skim the sheared-off edge of the plug. A machining head is provided to carry out these operations, and it can also be used if it is necessary to remove a scratch in the bore of the hole to be plugged or if it is necessary to remove a defective plug.

8

OTHER APPLICATIONS OF EXPLOSIVE WELDING

The main application of explosive welding is to flat plate cladding and the welding of cylindrical surfaces such as tube to tube and tube to tubeplate welding. In general explosive welding is not applicable to complex shapes, not only because of the difficulty of laying the explosive charge or the problem of providing adequate support, but because of the high tensile stresses generated by the reflection of the compressive waves from changes of section which will lead to failure. However, there are a few other applications which have been investigated, some of which have found limited industrial application and others which may have potential in the future.

An early application of explosive welding was the lap welding of metal sheets by using line welding and this has found some commercial application. Line welding can be used for fabricating large structures and it has been suggested for the construction of large structures in outer space. Scarfed weld joints can be used to produce a butt weld, and it can also be used to produce tubular transition joints. Line welding can be used to fabricate hollow sections, but this is also possible by cladding a recessed parent plate. Thin metal foil honeycombs can be produced by explosive welding. Indoors explosive welding for serial production has been used for joining copper sheathed aluminium wire coils prior to drawing down to smaller gauges, and for multiple spot welding of thin plates to form composite plate. Another application of potential promise is the explosive fabrication of wire reinforced composites.

The first mention of lap welding of sheets and its application to the fabrication of components was made by Holtzman and Rudershausen (134), but they failed to give any details of the technique to be adopted. Addison (186) was perhaps the first to report on the methods to be adopted to achieve a lap weld between metal sheets by use of a line charge, either in contact with the upper sheet or with a constant clearance. Using Primacord, a high detonation velocity explosive cord, he has adopted the inclined and parallel arrangements shown in Fig. 8.1. No advantage was found with the inclined arrangements, and the strength of the lap joint produced by the

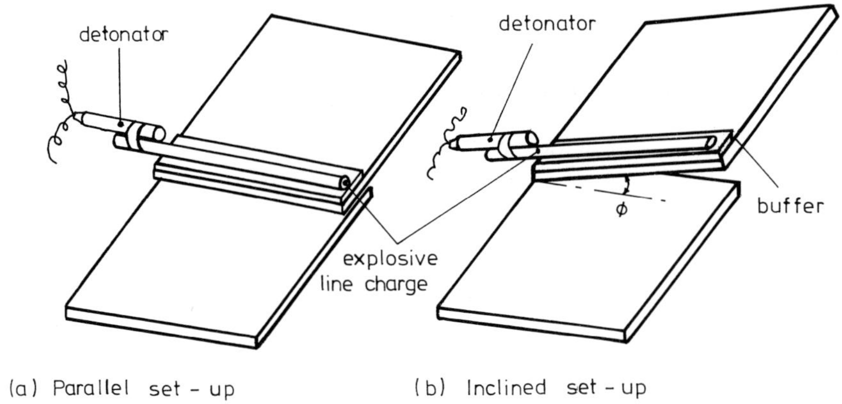

FIG. 8.1 Welding of lap joints (186)

parallel set-up was up to the parent strength. Kogya and Kaisha (187) have patented the lap welding arrangement of Fig. 8.2 with various charge configurations shown in the figure.

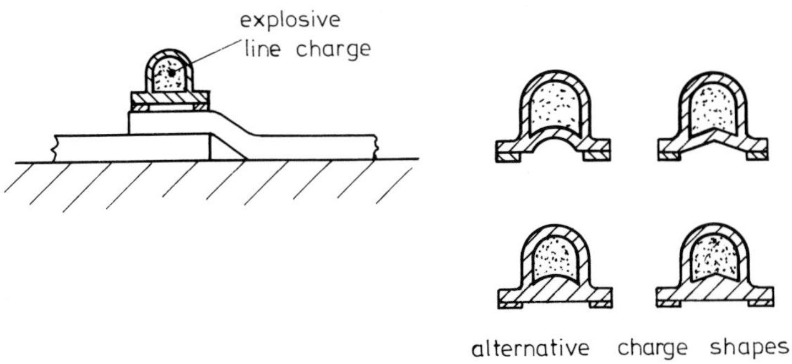

FIG. 8.2 Welding of lap joints (187)

Polhemus (188) noted various configurations for lap welding, but in particular the scarfed weld joint of Fig. 8.3. Shribman

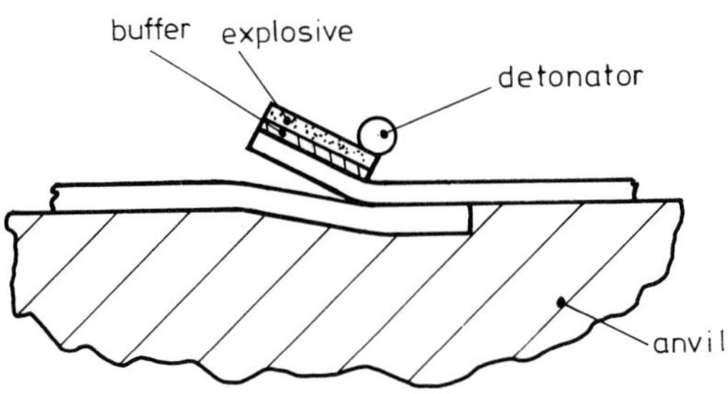

FIG. 8.3 Scarfed lap weld (188)

et al (88) have used the various line charges shown in Fig. 8.4

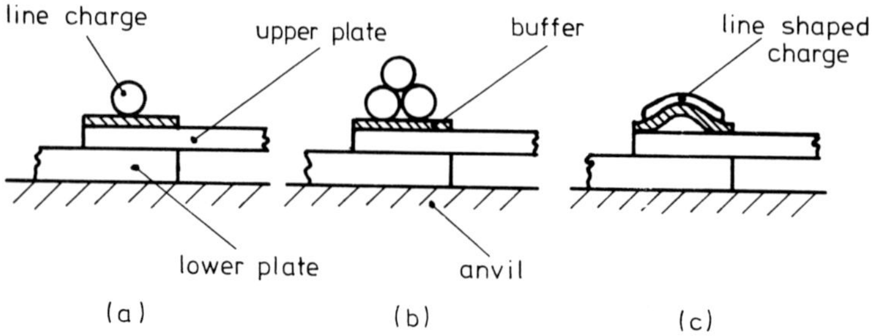

FIG. 8.4 Welding of lap joints (88)

to weld 0.8 mm and 1.6 mm brass sheets. The 0.8 mm sheets could be welded with a single run of Cordtex explosive cord, while the 1.6 mm sheets needed three runs of Cordtex cord but if there had been a positive stand-off between the two sheets better results might have been obtained. In tensile tests parent metal strength was achieved and failure occurred away from the lap joint. Metallurgical examination of the welds showed as might be expected, that there were two zones of welding one each side of the line charge as shown in Fig. 8.5, and

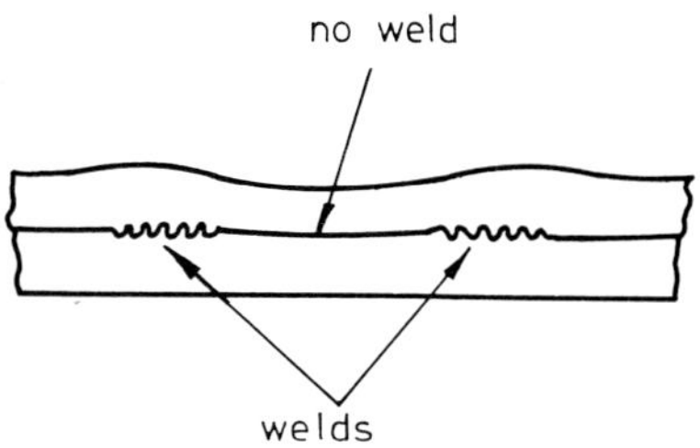

FIG. 8.5 Cross-section of lap weld (88)

the interfacial waves were normal to and running from the centreline of the explosive charge. Bement (189) used an explosive charge of RDX encased in a lead sheath that has been shaped into a thin rectangular cross-section, to form an explosive ribbon, which he has used with the configurations of Fig. 8.6 to form similar metal welds in aluminium alloy and titanium.

Lap welding can be applied to welding of straps joining two plates, tee joints, corner joints, longitudinal and circumferential welds in cylindrical members. Using these techniques

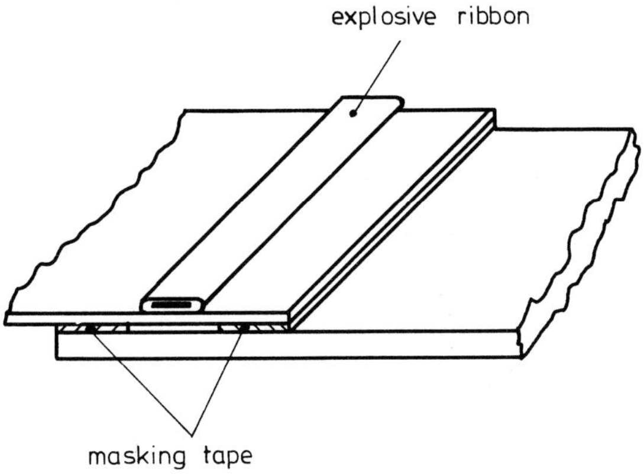

FIG. 8.6 Welding of lap joints (189)

Holtzman and Rudershausen (134) have indicated how the heat exchanger of Fig. 8.7 can be fabricated. The arrangements pro-

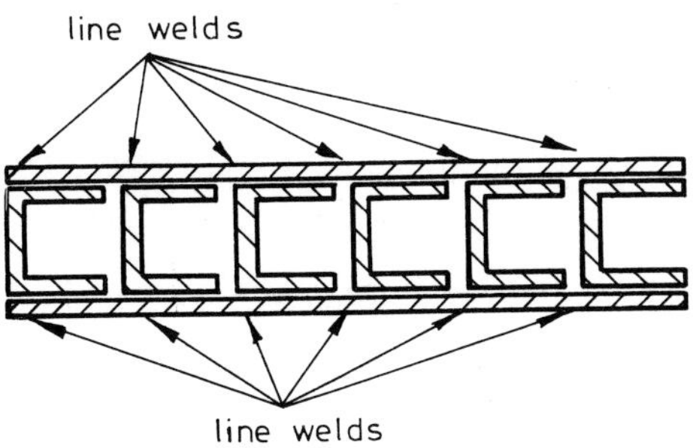

FIG. 8.7 Explosive fabrication of heat exchanger (134)

posed by Kogya and Kaisha (187) have been used to fabricate the large stainless steel lined steel water tank mentioned by Fukuyama (190). Polhemus (188) has mentioned the fabrication of cylinders, propellant supply tanks, spherical vessels and refractory metal combustion chambers including welds up to 10 m in length produced with a single detonation. Bement (189) has used explosive ribbon to fabricate a one twelfth scale model of a space station. Lastly Willis (191) has reported the successful line welding of aluminium foil to form a continuous area of 1600 m^2, which involved a total of 4500 m of welding with welds of over 10 m in length made at each firing.

It has been noted that Polhemus (188) has used a scarfed weld joint as shown in Fig. 8.3, and if the excess material is removed then it gives what appears to be an inclined butt weld.

This basic technique has been applied by Unterlerchner and Prümmer (192) to the production of titanium/steel tubular transition joints with diameters from 35 to 200 mm. Wright and Boyce (193) suggested the use of a scarfed weld joint of the form shown in Fig. 8.8a to produce tubular transition joints, and Grollo (152) has used the similar arrangement of Fig. 8.8b to produce 63.5 mm and 101.6 mm diameter tubular

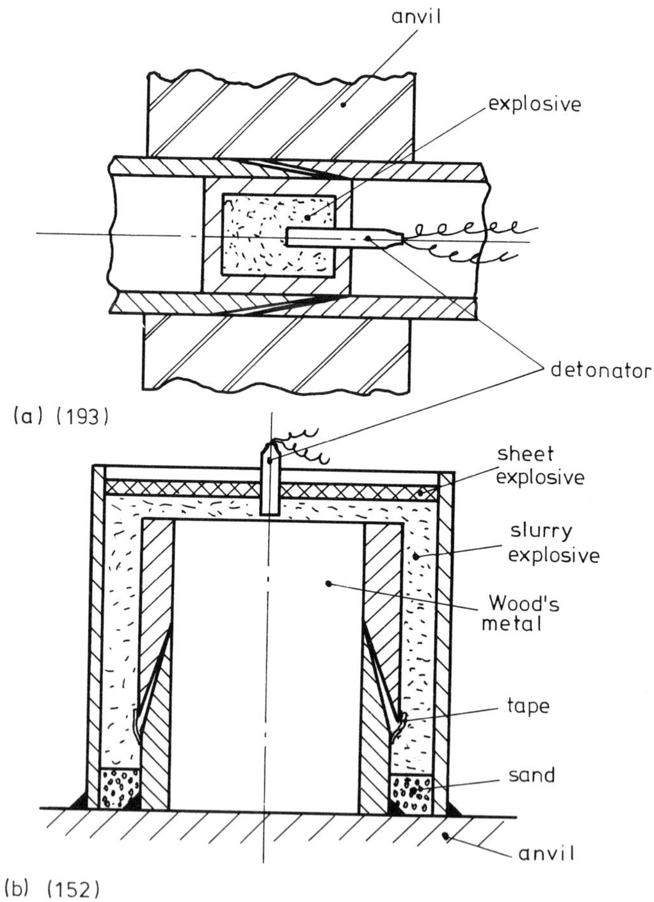

FIG. 8.8 Explosive fabrication of transition joints, (a) (193), (b) (152)

stainless steel to aluminium transition joints. As will be seen the two tapers have a slightly different angle to provide a stand-off between the two components. After welding the bore and the outer diameter must be machined to remove the inevitable crevices. Tubular transition joints are also machined out of cladded plate; for instance Izuma and Baba (194) have developed a five component cladded plate for the manufacture of a stainless steel to aluminium alloy tubular transition joint for cryogenic temperatures.

The scarfed type weld joint can be used for joining two sheets as illustrated in Fig. 8.9a. There is also reference by Holtzman and Cowan (195) to the possibility of producing a straightforward butt weld using the set-up of Fig. 8.9b, but a substantial charge would be required to cause lateral plastic flow, and there will be a reduction of thickness of the sheets under these explosive charges. It is probable that crevices will be left top and bottom, which could be a source of weakness unless they are removed.

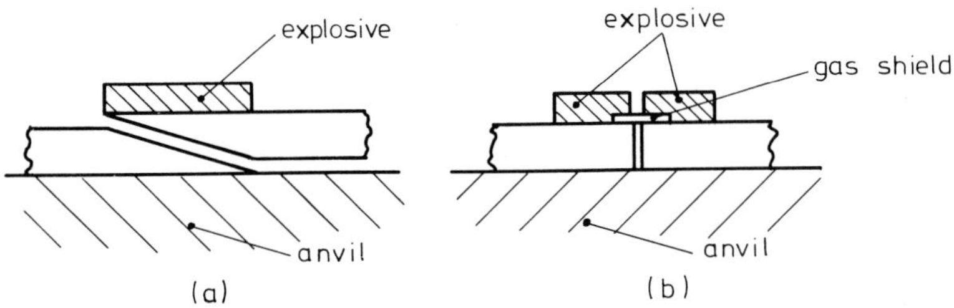

FIG. 8.9 Butt welding (195)

Hollow sections can be fabricated by line welding as illustrated by Fig. 8.7, but there is an alternative way of producing hollow sections. For instance a hollow section such as that illustrated in Fig. 8.10 can be fabricated by machining recesses in the parent plate and explosively welding the cover plate to the parent plate. However, the kinetic energy is so high that the cover plate would be sheared-off on the edges of the recess. This can be overcome by filling the recesses with a low melting point alloy, and then welding the cover plate to the parent plate and subsequently melting out the low melting point alloy. There is the danger that the low melting point alloy in the jet could contaminate the weld interface between the cover and parent plates, and to minimise this it is necessary to relieve the low melting point infill in the area adjoining the edges of the recess.

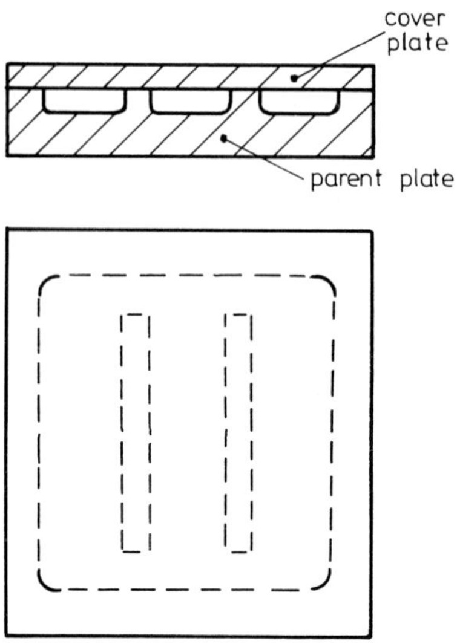

FIG. 8.10 Hollow section which can be fabricated explosively

Wylie and Crossland (196) have produced tubes with two rows of leak detection channels incorporated in the wall thickness, as a feasibility study for a liquid metal/steam heat exchanger. This was achieved by the implosive welding of three tubes as illustrated in Fig. 8.11, the inner two tubes of which

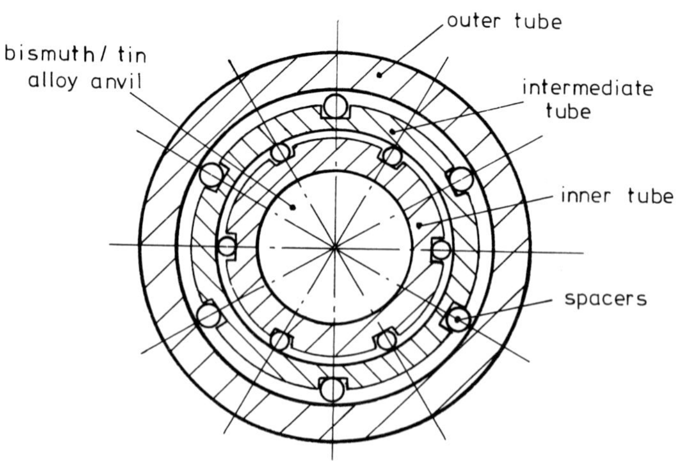

FIG. 8.11 Explosive fabrication of triplex tubing with leak detection channels (196)

had axial grooves machined in them. Prevention of the collapse of the grooves and maintaining the initial clearance between the three component tubes were achieved by inserting a filler filament in the grooves. Aluminium and lead wires, nylon filament and PVC tubing filled with water have been used, and from the few tests completed water filled PVC tubing seemed to provide the best solution. The inner cylinder was supported to prevent it collapsing by filling it with a bismuth-tin eutectic alloy which expands on solidification, and this appeared to be entirely adequate. Though the study demonstrated the feasibility of producing such tubes further development would be needed to extend to long lengths of tubes.

Honeycombs can be produced by explosive welding. Wright and Bayce (193) have shown photographs of a copper honeycomb produced by welding together a bundle of copper plated aluminium wire held in a copper tube, which was subjected to an implosive explosive charge. The aluminium was then dissolved out leaving a copper honeycomb.

Velten (197) and Yoblin and Mote (146) have both described a commercially more attractive method of fabricating honeycombs. Metal foil is printed to stop welding in certain areas, and they are then stacked so that the printed area alternates from sheet to sheet as indicated in Fig. 8.12a. The pack of foil is

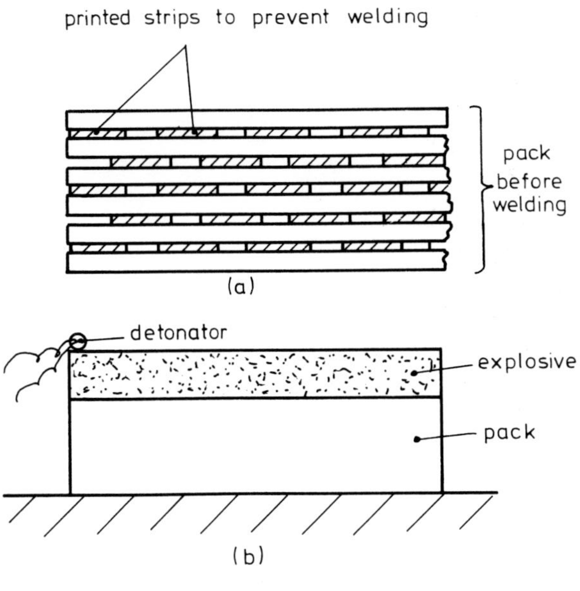

FIG. 8.12 Explosive fabrication of honeycombs (146)

then explosively welded with a low detonation velocity explosive as shown in Fig. 8.12b. It is then opened out to form the honeycomb. Honeycombs with a 4.75 mm cell size made from 25 to 50 μm thickness foil have been produced in a range of materials including titanium, stainless steel, aluminium alloy, hastelloy, nickel, inconel, etc. In a single operation it is possible to produce an expanded honeycomb 2.5 x 1.25 x 0.625 m.

Indoor explosive welding for serial production has been employed in Sweden and Japan. Persson (198) has described the explosive joining of copper cladded aluminium wire. In this process a billet of aluminium sleeved with copper is hydrostatically extruded to rod, and then it is drawn down to the final wire size required. However, the hydrostatic extruder can only accommodate 100 kg of material whereas the final wire drawing is a continuous process, so it is necessary to join the coils of rod prior to drawing. This is carried out explosively by the method illustrated in Fig. 8.13, but it is necessary to

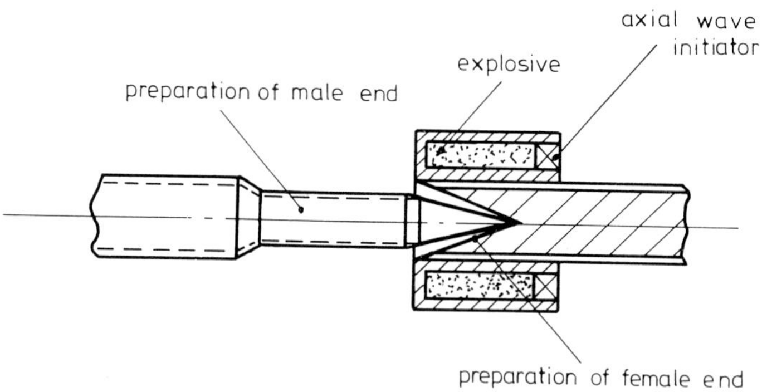

FIG. 8.13 Explosive joining of copper cladded aluminium wire (198)

carry out this process as part of the line production system. After machining the two ends of the successive coils the explosive weld is carried out inside an explosion chamber in the factory. Persson also describes a plant for explosively spot welding bimetal plates. He notes that explosive cladding of very big and thin plates is costly, and frequently it is not necessary to have a 100% bond and two plates which were spot welded together would be adequate. Consequently a machine has been developed through which two plates are fed in steps into an enclosure, where they are clamped together and a line of small explosive charges are detonated simultaneously to create a row of spot welds. This is repeated in regular steps with a cycle time of 60s per step.

Kameishi et al (199) have briefly described the explosive cladding of cast steel disc slides of gate valves of 150 to 600 mm diameter with a ring of 2.5 - 2.8 mm thick martensitic stainless steel or stellite. The demand for the 150 mm valve is such that an automatic cladding machine has been developed in which the explosive charges and flyer plates are automatically placed, and the disc valve and charges automatically transferred to an explosive chamber in which the charges are simultaneously detonated.

Persson (198) has briefly mentioned explosive spot welding for fabricating composite plate where a continuous weld is not required. Experiments on explosive spot welding has been briefly noted by Addison (186) and Wright and Bayce (193), and basically Carl (4) observed explosive spot welding when he was the first to note explosive welding. Basically an explosive spot weld is similar in concept to a line weld; a small cylindrical charge or detonator placed normal to two thin plates held in contact causes an oblique impact which runs radially from the centre of impact between the two plates. This leads to an annular area of welding with an area of no bond at the centre as observed by Carl. It is possible to spot weld similar and dissimilar metals and this would be of commercial interest if a simple, repetitive and cheap method could be devised for placing the small charges required. An extremely interesting possibility has been noted by Erdmann-Jesnitzer et al (200) who observed that a spot weld between thin metal sheets is caused by the impact of a high velocity slug of water. This spot weld has similar characteristics to an explosive spot weld. An alternative solution to spot welding thin sheets would be to devise a method of generating high velocity slugs of water.

It is well known, because of the existence of imperfections or dislocations within the structure of bulk solids, that the actual strength of most materials is much less than the theoretical strength. Whiskers of some solids such as iron and carbon have been produced which are free of dislocations, and they exhibit strengths of the order of their theoretical values. However, the small diameter and length and the difficulty of production of such whiskers has prevented their exploitation in engineering. Fibres of some materials, notably glass and carbon, have given strengths which begin to approach those of whiskers. Also the strength of fine metal wires increase as their cross-sectional area decreases and strength levels of three times that of bulk material can be achieved.

Fibres and thin metal wires are of limited use by themselves as their high strength cannot be utilized; however, they can be usefully employed if they are incorporated in a tough and ductile matrix. The matrix material separates, protects and aligns the fibres or wires to form composite materials of high strength and good fracture properties which have great engineering potential. For example glass or to a growing extent carbon fibre in a resin mixture is now extensively used in the fabrication of shells. Many attempts have been made to produce composites of high strength fibres and wires in a metal matrix.

Wire reinforced metal matrix composite materials have been made by powder metallurgy techniques, infiltration of wires by molten matrix metal, hot rolling of wire between sheets and diffusion bonding by hot pressing of wires between metal sheets. All these methods involve the use of high temperature which can cause softening of cold drawn wires, and wire/matrix interactions that produce brittle intermetallic compounds and surface defects in the wires. As a result, the tensile strength of the composites may be reduced. In addition these methods of production are subject to practical limitations in the size of composite plate that can be produced.

Jarvis and Slate (201) were the first to publish a paper on the explosive fabrication of fibre reinforced metals. They produced specimens with tungsten wire incorporated in a copper matrix, which was fabricated from 0.3 mm thick layers of copper foil interspersed with 0.15 mm diameter tungsten wire to give a volume fraction of 17 per cent. Though they adopted a parallel configuration and used a high detonation velocity explosive which would not normally permit welding, nevertheless with wires between the foils perfectly adequate welding between adjoining foils was observed, and there appeared to be a fairly good bond between the wires and the matrix. The ultimate strength of the composite was found to be that predicted from the strength of the two starting materials and their volume fraction. Slate and Jarvis (202) reported further work on tungsten wire in a copper or aluminium matrix, and beryllium wire in an aluminium matrix in which a 38 per cent volume fraction was achieved. Steroscan electron microscope examination indicated regions of adhesion between the matrix material and the wire. These areas of adhesion were seen to be mainly in regions of the wire surface where there is a non-uniformity of the surface or where the geometry of the impact between the foil and the wire favoured welding.

Fleck et al (203) briefly reported on wire reinforced aluminium produced by a parallel set-up using a low detonation velocity explosive. From transverse bend tests they concluded that there was little or no fibre to matrix bond, so they heat treated the composite to promote bonding. Tests on this heat treated material gave an ultimate tensile strength which was in agreement with that predicted from the volume fractions.

Reece (204) (205) has reported on TZM molybdenum wire reinforced C129Y columbium, which has a potential application to the construction of the space shuttle. He used a parallel set-up with nitroguanidine, a low detonation velocity explosive, to weld columbium sheets together with an interspersed layer of molybdenum wires. If thicker sheets were required a single layer of molybdenum wires and a columbium sheet were welded to each side of the original sheet. Photomicrographs showed considerable cracking and deformation of the molybdenum wires, but mechanical tests indicated an ultimate tensile strength of the composite which was appreciably higher than that predicted by the law of mixtures.

Wylie et al (206) reported on the explosive fabrication of high tensile steel wire reinforced aluminium. They used a parallel set-up with one layer of wires stretched in the stand-off between the two plates, together with a low detonation velocity explosive, Trimonite 1. Different volume fractions were achieved by varying the thickness of the two aluminium sheets, though in a much larger specimen the volume fraction was varied across the plate by varying the pitch of the wires. In the elastic region the modulus of the composite according to Kelly (207) is given by:

$$E_c = E_f V_f + E_m(1 - V_f) \tag{8.1}$$

where,

E_c is the Young's modulus of the composite
E_f is the Young's modulus of the fibre
E_m is the Young's modulus of the matrix material
V_f is the volume fraction of uni-directional fibres

The experimental data gave a considerable scatter of values of modulus, and no reasonable comparison with predicted values was apparent. However, at higher strains where the matrix has yielded while the fibres remain elastic the effective modulus according to Kelly is given by:

$$\left(\frac{d\sigma_c}{d\varepsilon_c}\right)_{\varepsilon_1} = E_f V_f + \left(\frac{d\sigma_m}{d\varepsilon_m}\right)_{\varepsilon_1} \tag{8.2}$$

where,

$\left(\frac{d\sigma_c}{d\varepsilon_c}\right)_{\varepsilon_1}$ and $\left(\frac{d\sigma_m}{d\varepsilon_m}\right)_{\varepsilon_1}$ are the respective slope of the stress/strain curves for the composite and matrix at a strain of ε_1.

At a strain of 0.35% the measured modulus of the composite appeared to be somewhat greater than that predicted by Equation 8.2.

The maximum strength of the composite according to Kelly and Tyson (208) depends on the relative properties of the fibres and the matrix material. If it is assumed that both materials fail at small strains before the onset of plastic instability, and the strain to failure of the matrix material is greater than in the fibre material, then the variation of strength will follow curve AOB in Fig. 8.14. At small volume fractions below the critical value V_f' the fibres will have failed before the matrix material has reached its maximum strength, so the com-

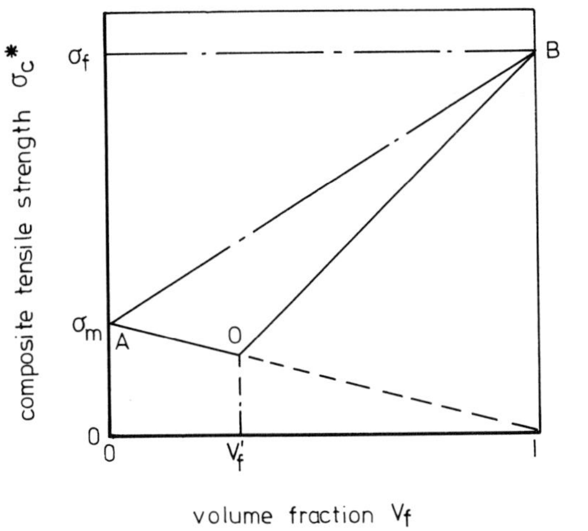

FIG. 8.14 Idealised variation of composite tensile strength with volume fraction of reinforcement

posite acts as though it was full of holes with its true area reduced by the volume fraction. Above the critical volume fraction the maximum strength is dictated by the fibre material, and it is reached at a strain equal to that at the maximum stress in the fibre material. This composite gives the following maximum strength:

$$\text{for } 0 < V_f < V'_f,\ \sigma_c^* = \sigma_m^*(1 - V_f) \qquad (8.3)$$

$$\text{for } V_f' < V_f < 1,\ \sigma_c^* = \sigma_m(1 - V_f) + \sigma_f^* V_f \qquad (8.4)$$

where,

σ_c^* is the tensile strength of the composite

σ_m^* is the tensile strength of the matrix material alone

σ_f^* is the tensile strength of the fibre material alone

σ_m is the stress in the matrix material at the strain to failure in the fibres

V_f' is the critical volume fraction at which the strength of the composite material is a minimum

However, if the strain to give the maximum stress in both the matrix material and the fibres is the same, then the variation of strength with the volume fraction follows the 'law of mixtures' which can be simply stated as:

$$\sigma_c^* = \sigma_m^*(1 - V_f) + \sigma_f^* V_f \qquad (8.5)$$

for all values of V_f. This is the line AB in Fig. 8.14.

The tensile strength of the composites were also established by Wylie et al (206), and this corresponded well with values predicted by Equation 8.5. The tensile strength of the aluminium was determined from explosively welded material, though the properties were not significantly changed by shock loading. However, corresponding data on shock loaded wire could not be established so properties in the as received state were used. They also noted that excessive wave amplitude at the interface caused serious damage to the wires.

Wylie (209) also explored the possibility of achieving higher volume fractions by using 48 mesh stainless steel mesh having a wire diameter of 0.2 mm between soft aluminium sheet 0.6 mm thick. A stack of seven layers of mesh and eight layers of aluminium were welded in a single operation and a volume fraction of 0.16 was achieved, though this is equivalent to 0.08 in each of the principal directions. The limited tensile tests carried out on this material indicated a tensile strength much less than that predicted by Equation 8.5, which was attributed to the severe kinking and overlapping of wire always present in mesh. Wylie also carried out a few exploratory tests with carbon fibres but with no success; the carbon fibres were fractured and appeared to act as an excellent lubricant between the two plates.

Bhalla and Williams (210) have reported a considerable extension of the work of Wylie et al. For their initial work they used the parallel set-up to repeat the work on the fabrication of composites with multiple layers of stainless steel mesh with intervening soft aluminium sheets. They concluded that the mechanism of welding in this application is not the same as for conventional explosive welding, and they suggest that aluminium is extruded through the gaps in the mesh. It is the flowing together of newly exposed clean aluminium which produces the bond, as in the case of cold pressure welding. This thesis is supported by the experimental set-up shown in Fig. 8.15 in which it will be noted that the detonator appears to be at the wrong end of the explosive charge, but in fact it gives a normal impact between the aluminium driver plate and the stack of wire mesh and foils if $V_P/V_D = \sin\alpha$. With this arrangement no welding would occur in the absence of meshes,

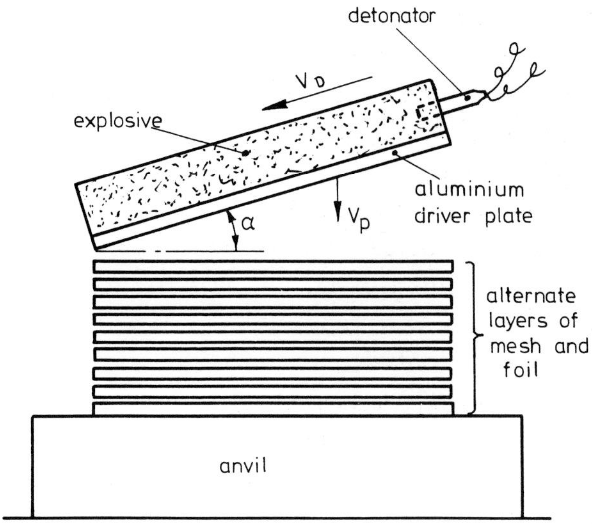

FIG. 8.15 Explosive welding of mesh and foil under normal impact (210)

whereas in their presence well bonded composites were produced. Further evidence of the effect of the mesh is provided by the results of experiments with stacks containing six or seven layers of mesh, in which the size of the apertures in the mesh was varied while keeping the driver plate kinetic energy in the range 120 to 220 J/cm^2. It was found for a mesh of 40 x 40 (wires/in) with a wire diameter of 0.254 mm which gives an aperture of 0.381 mm that welding occurred, whereas for a mesh of 60 x 60 with a wire diameter of 0.193 mm which gives an aperture of 0.23 mm no welding occurred. With non-isostropic meshes welding was found to occur as long as one dimension of the mesh aperture was 0.381 mm or greater. It appears for this kinetic energy that the pressure generated is insufficient to allow the aluminium to flow through a mesh which is smaller than 0.381 mm.

Bhalla and Williams concluded that with the energy levels they used there was no bonding between the aluminium and the steel wires, and this is supported by examination of the tensile failure of the composite. Fig. 8.16 shows the microstructure of the isotropic mesh and the tensile fracture surface.

Perhaps the main contribution of Bhalla and Williams is their work with meshes in which the warp was high tensile stainless steel wire and the weft was of soft stainless steel or

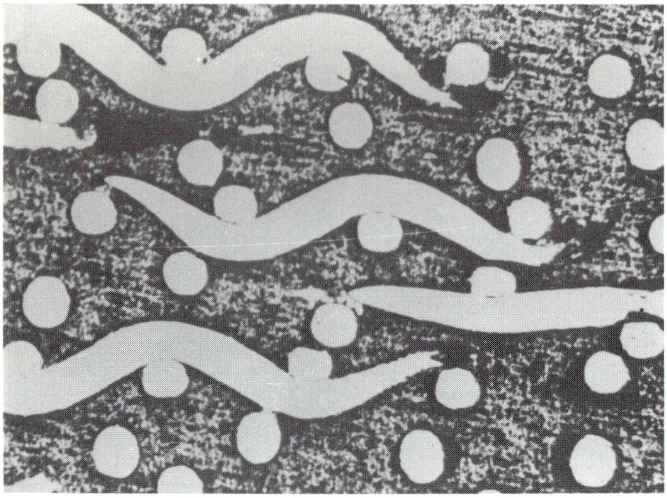

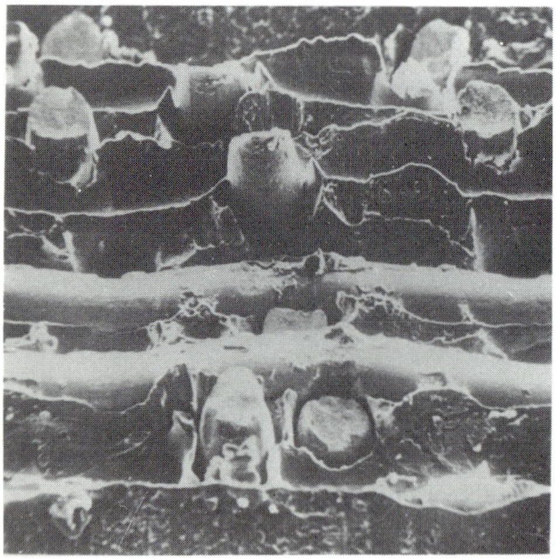

FIG. 8.16 Steel mesh/aluminium composite (210). Mesh 40 x 40, 0.254 mm diameter soft stainless steel, foil 0.254 mm cold rolled pure aluminium. (a) microstructure x 40, (b) tensile fracture surface x 60

aluminium wire. Not only was the material of the weft different to that of the warp but also the mesh was much less for the weft than it was for the warp. The weft served to hold the high tensile wires in position during the welding operation. A composite with a mesh of 50 x 10 stainless steel/aluminium with 0.254 mm diameter high tensile stainless steel wire and 0.254 mm thick aluminium foil gave a volume fraction of 0.24 in the principal direction with no fracturing of the wire. Tensile tests on this material gave a tensile strength of 97% of the value predicted by Equation 8.4, and almost six times the tensile strength of the aluminium matrix. Fatigue tests in repeated tension on this material gave a fatigue stress at 10^7 of 1.43 times the tensile strength of the matrix material, and four times the fatigue stress at 10^7 cycles in the matrix material. Metallographic examination of the region near the fracture showed that in some instances fatigue cracks had propagated continuously through the matrix and wires. Fig. 8.17 shows the microstructure of the high tensile stainless steel/ aluminium mesh and the fatigue fracture surface.

Bhalla and Williams also demonstrated a possible technique for producing large areas of composite. The problem is a practical one of how to hold the stack flat when the mesh is springy and the foil not flat. They resolved this by placing the stack in a polythene bag and then evacuating the air with a rotary pump to give a coarse vacuum of 15 mm Hg, which was held while the charge was detonated external to the bag. Atmospheric pressure was sufficient to compress the stack. Using this technique they produced composite plates with dimensions 300 x 500 mm with a volume fraction of high tensile stainless steel wires of 0.24. They concluded that there was no reason why larger sheets could not be produced using this arrangement.

The feasibility of fabricating wire matrix reinforced cylindrical elements has been studied by Blazynski and El-Sobky (211) (212) (213). They have examined various configurations involving stainless steel wire mesh in aluminium, copper, brass and steel matrix material, and they state that satisfactory implosive welding of complex composite cylinders can be achieved.

Explosive swaging for joining electric power transmission cables and for the attachment of end fittings was described by Haegland (214). Willis (191) has used explosive swaging of the compression fittings for 400 kV down-dropper fittings, and he also considered that this technique could be applied to overhead line terminations to reduce interface resistance and allow transmission line uprating. Blazynski and Bedroud (215) have reported on the implosive welding of co-axial multi-strand composites, which they envisage as having application in the

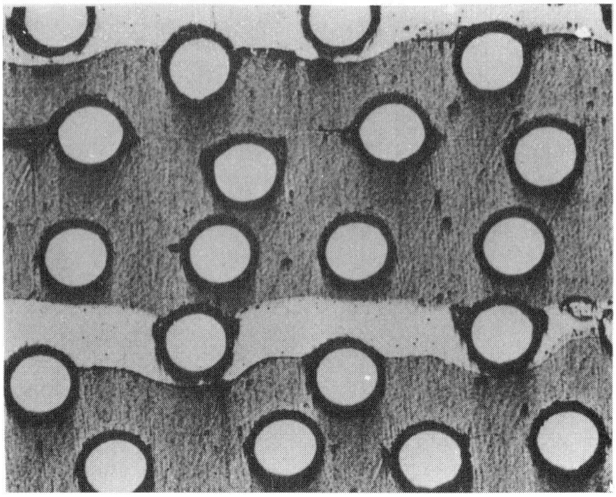

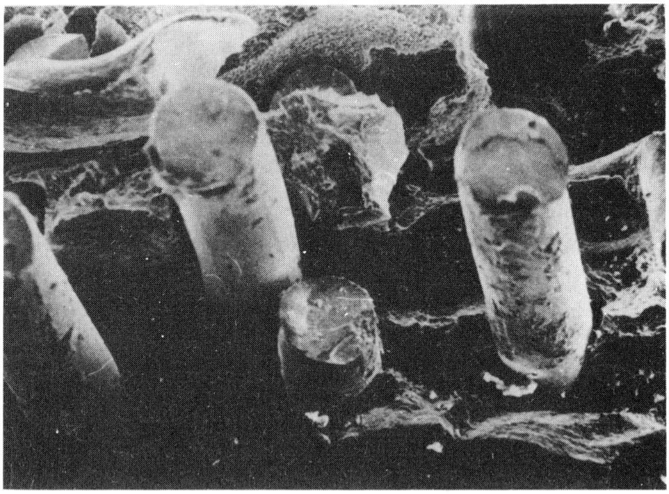

FIG. 8.17 Steel mesh/aluminium composite (210). Mesh 50 x 10, 0.254 mm diameter high tensile stainless steel by 0.274 mm diameter aluminium. Foil 0.254 mm cold rolled pure aluminium. (a) microstructure x 50, (b) fatigue fracture surface x 60

atomic energy and chemical industries, either in the as welded state or after a suitable metal-forming operation such as extrusion. They have welded various arrangements of rods sleeved in metal or plastic including various combinations and permutations of brass, copper, mild steel, stainless steel and titanium.

9

METALLURGY OF EXPLOSIVE WELDING

In a perfect weld the two components are joined at a common interface such that no evidence of the weld remains. Such a weld would result if perfectly mating surfaces of two similar crystals were brought together to within a distance where the interatomic forces became effective. The surfaces would then approach to the interatomic equilibrium distance, where the interatomic attractive and repulsive forces are in equilibrium or the potential energy is a minimum. Thus two single crystals would become one, and not even a grain boundary would remain to indicate that a weld had been made.

Normally there is some evidence of the weld, as in explosive welding. In its simplest form the weld zone would be a grain boundary. This would be the case when perfectly mating surfaces of two misoriented single crystals or two polycrystalline components are brought sufficiently close together for bonding to occur. It has frequently been stated in theoretical discussions on solid phase welding that large misorientations between crystals result in weaker welds than those formed between similarly oriented crystals. Such statements are only valid in the case of the bonding of perfect crystals, whereas the presence of imperfections reduce the yield strength of metallic crystals by a factor of about one thousand, so that even the worst misoriented grain boundary does not yield before yield occurs in the adjacent crystals. If this was not so it would be common to all polycrystalline metals to fail in an intergrannular fashion, and this is not normally found to be the case.

In a pressure weld welding is achieved by causing gross plastic flow at the interface between the components, which ruptures the surface film and allows fresh metals surfaces to come into contact. In such welds the interface consists of a grain boundary containing the fragments of the original surface films, while the zones adjoining the interface may be hot or cold worked depending on whether the pressure weld has been carried out cold as in cold pressure welding, or hot as in forge welding. It is possible for some limited solid state interdiffusion to occur across the interface. The essential problem of pressure welding is if the surface films are not sufficiently brittle to rupture easily, or if the plastic flow characteristics of the two metals are very different. In either case excessive deformation will be required to achieve welding, which is probably unacceptable.

In explosive welding the surfaces of the two components being joined are subject to severe disruption by jetting, which removes the adjoining surface layers, and by the interfacial waves which are frequently observed at the weld interface. Explosive welding has similarities to pressure welding in that the two surfaces are subject to gross plastic flow, but there are three major differences. Firstly, in explosive welding the gross plastic deformation occurs in a very localised region adjoining the weld interface, whereas in pressure welding both components are subject to gross plastic deformation. Secondly, in explosive welding both components suffer the passage of shock waves, which may be sufficiently intense as to cause structural changes, and which causes a significant increase of hardness of the components with a resulting reduction of impact strength and fracture toughness. If the shock wave is sufficiently intense then the reflected tension wave may exceed the tensile strength of the materials causing what is termed a spalling failure. Such failures may in particular occur in the flyer plate, which is subjected to the shock wave generated by the detonating explosive. Thirdly, in the vortex regions where part of the considerable kinetic energy of the re-entrant jet is dissipated there is frequently evidence of melting such as a cast structure with small cooling cavities. These small molten zones contain a mixture of the two component metals, and they are subjected to very high cooling rates which may result in the forming of metastable compounds.

An understanding of the metallurgy of explosive welds is important in the overall assessment of their mechanical properties. Obviously to be able to relate metallographic examination of explosive welds to their mechanical properties it is necessary to be able to measure these properties.

The mechanical testing of a few particular composites, chromium steel/steel, stainless chromium-nickel steel/steel, nickel and nickel-base alloy/steel and copper and copper alloy/steel, is covered by American National Standards (216-219). However, though the mechanical tests required by these standards are no doubt adequate as acceptance standards they are not sufficient to fully describe the mechanical properties of the bond. The essential problem in the mechanical testing of explosive welds is that the cladding is mostly thin, and consequently it is difficult to devise a test to measure the strength of the bond under simple and well defined stress states such as simple tension or shear.

An obvious measure of the strength of the bond would be the tensile strength normal to the weld interface. Bahrani and Crossland (39) (220) have carried out tension tests by cladding a thick block of metal, and then machining a tension specimen from it as shown in Fig. 9.1. They reported results

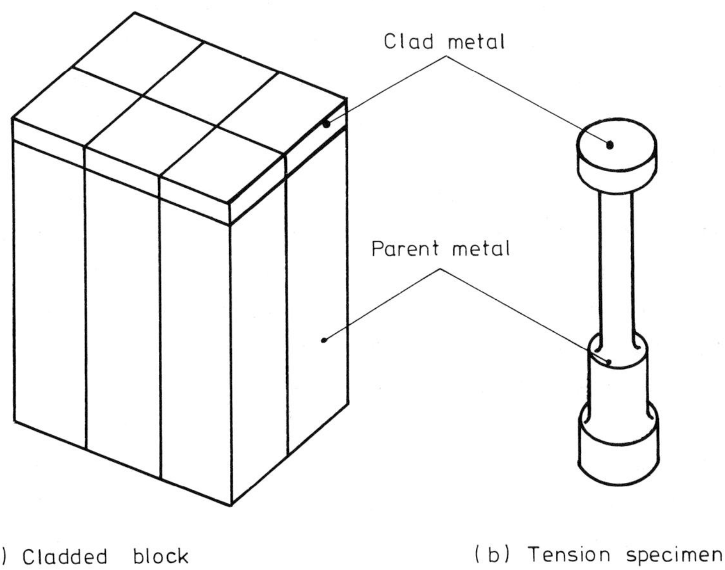

FIG. 9.1 Tension test piece normal to weld interface (39, 220)

for 3 mm stainless steel and titanium welded to mild steel both for the as welded and annealed states. They noted that weld strengths in excess of the strength of the mild steel could be achieved with stainless steel to steel welds, with a typical necking failure removed from the weld zone. With titanium to steel welds a high strength was measured in the as welded state which was seriously reduced by annealing at 850°C. This serious reduction was caused by the formation of a hard and brittle intermetallic alloy at the weld interface. It must be noted that the weld zone is adjoining the enlarged head of the specimen, and consequently the zone is much more constrained than material in the gauge length of a normal tensile specimen, and it is also subjected to a peeling action especially if the cladding is thin. Consequently the tensile stress measured cannot be directly compared with the ultimate strength of the two component materials. Though the data obtained are of value this form of tensile test is of little general applicability, as in most cases the cladded plate is not thick enough to allow the machining of such specimens.

Other forms of tension tests have been reported by Frey (221) and Anderson (222). Frey suggested the test specimen illustrated in Fig. 9.2 in which a bar of the appropriate metal

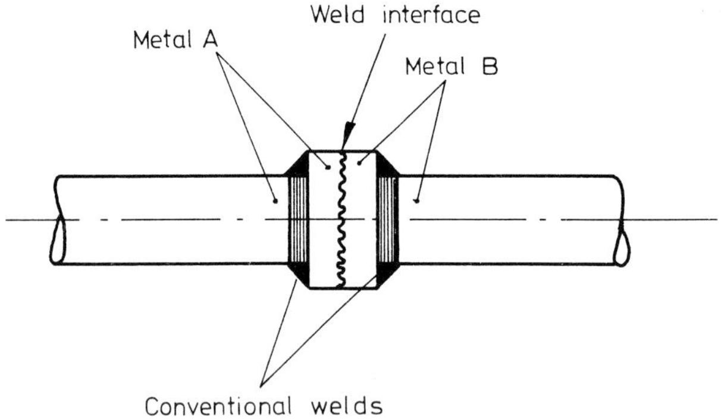

FIG. 9.2 Tension test piece normal to weld interface (221)

is fusion welded to each side of the explosively welded sample, so that a conventional tension test can be carried out across the explosively welded interface. However, the explosive bond will have been subjected to high temperature during the fusion welding, though this may simulate the effect of fusion welding which will normally be used in fabrication involving the use of explosively cladded plate. It should be noted that the area of explosive bonded sample is greater than that of the bars welded to it, so that the strength of the bond can only be determined if it is appreciably lower than the weaker of the two metals involved. An alternative test suggested by Frey and also reported on by Anderson is the ram tensile test illustrated in Fig. 9.3. Anderson comments on this test that though it ap-

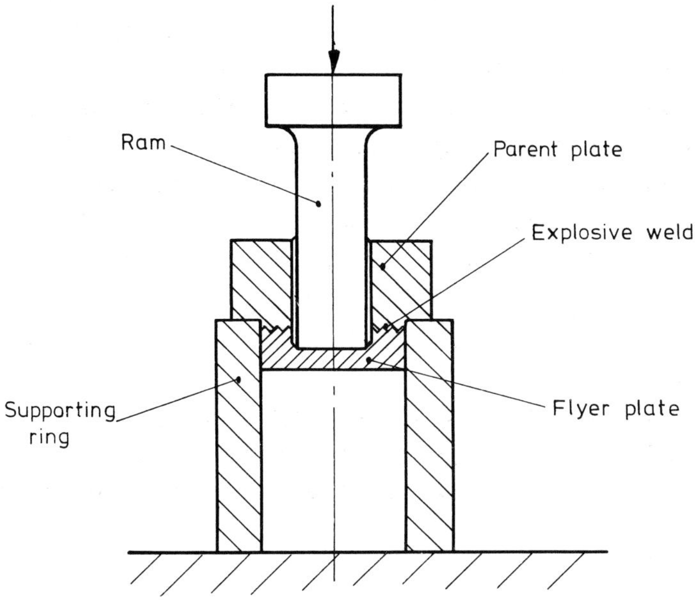

FIG. 9.3 Ram tensile test (221, 222)

pears to be a very sensible test it is only practical where there is a reasonable thickness and strength of the cladding material. The weld interface is much more constrained than in a conventional tension specimen, and additionally it is subjected to a peeling action in addition to the tensile stress. Consequently the tensile strength values cannot be related to the tensile strength of the component metals as obtained by a conventional tensile test piece.

Shear tests are commonly carried out. Fig. 9.4 shows a tensile shear test in which two saw cuts or milled slots are made one each side of the specimen down to a point beneath the weld interface. When the specimen is pulled in tension the area of the bond between the saw cuts, that is LW, is subject to shear. However, there is still a peeling action superimposed on the shear. A preferable shear test which is specified and detailed in the American National Standards (216-219) is the ASTM side shear test which is shown in Fig. 9.5. According to Anderson

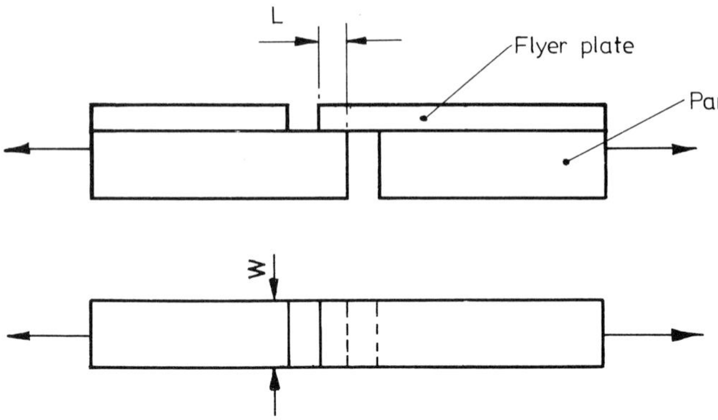

FIG. 9.4 Tensile shear test (39)

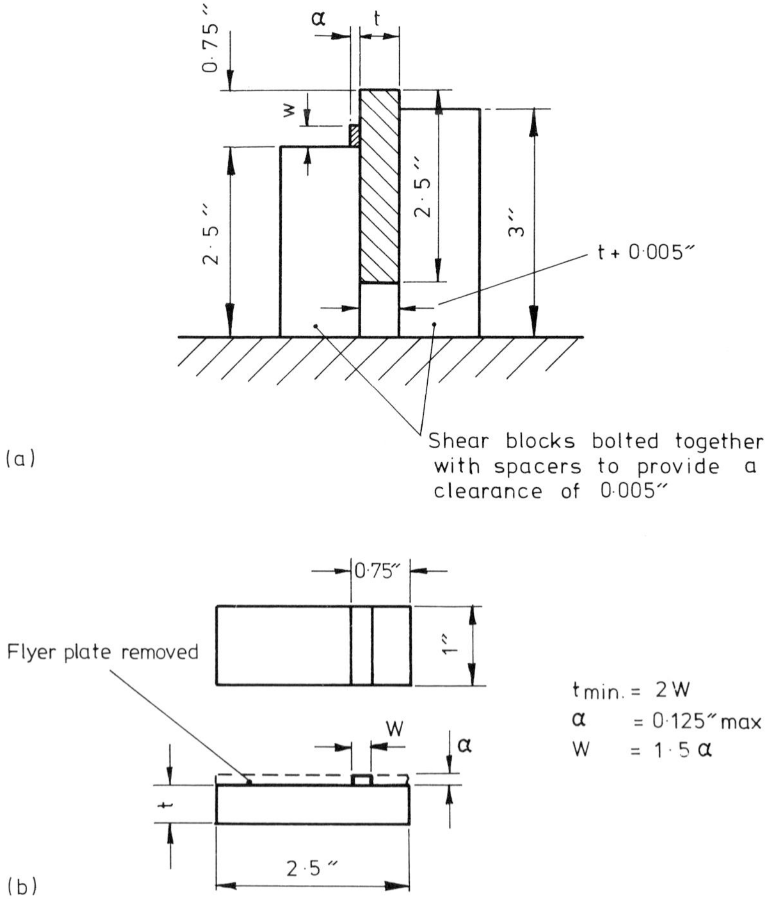

FIG. 9.5 ASTM side shear test (216-219)

(222) this test is less valuable than might be thought due to the inevitable interference of the interfacial waves during shear. The value of the shear stress depends upon the direction of shearing relative to the waves, and the depth of machining which defines the particular plane on which shearing will occur. If the depth of machining is too great then shearing will tend to occur in the parent plate material, or if it is not sufficiently deep then shearing will occur wholly in the flyer plate or cladded material. However, despite these criticisms the ASTM side shear test is probably the simplest and most reliable test which is available for testing the strength of the bond in explosive welding.

Qualitative tests such as the bend test of Fig. 9.6 are

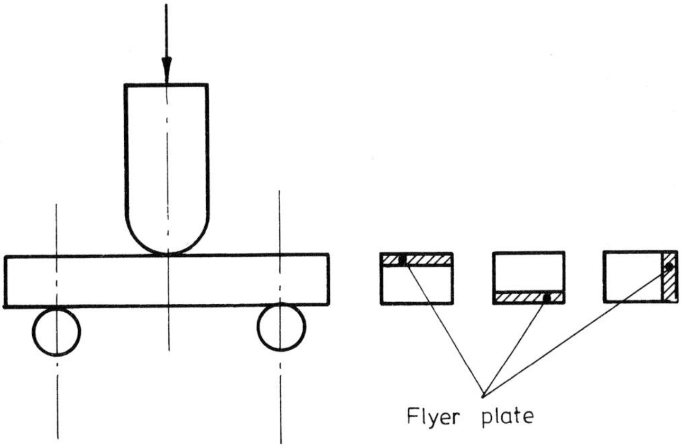

FIG. 9.6 Bend tests

frequently carried out, in which the cladding material is either on the inside or outside of the bend. Additionally a side bend test can be carried out in which the cladding material is on the side of the bend. Bend tests are included in the American National Standards (216-219). Another test is to get a square section strip of the cladded plate as shown in Fig. 9.7 and to twist it about its longitudinal axis in a torsion machine, which is a stringent test of the quality of the bond. An even simpler test is to use a hammer and chisel to try and separate the cladding from the parent plate. Though this may appear to be a rather crude test it is very effective if the bond is weak or if there is a brittle intermetallic compound at the interface

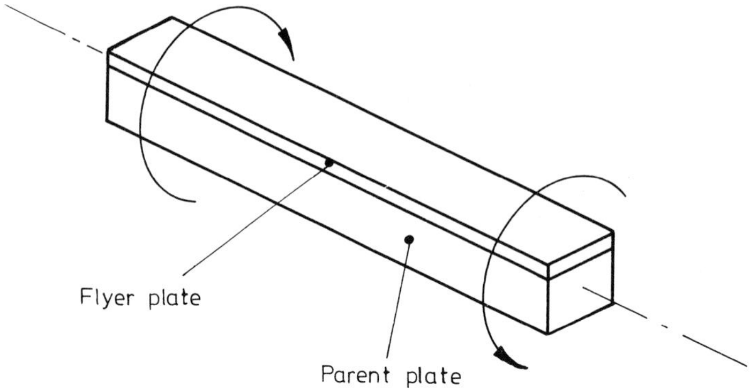

FIG. 9.7 Torsion test

because it is applying a shock load to the interface. The effectiveness of this test can be improved by machining a slot along the interface in which the edge of the chisel can be inserted as in Fig. 9.8. In the case of tube to tubeplate welds

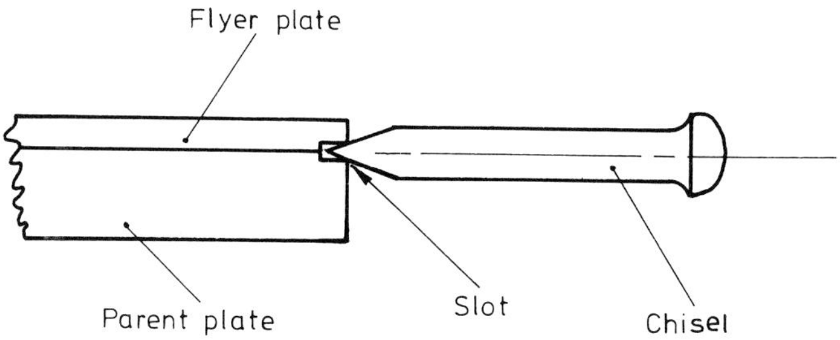

FIG. 9.8 Chisel test

in which the length of the weld is short, a narrow strip of tube is isolated by two radial cuts through into the tubeplate as in Fig. 9.9. An attempt is then made with a pair of pliers to peel the strip of the tube from the tubeplate.

The tests described so far are essentially static tests with the exception of the chisel test. However, the flyer and parent plates are hardened by the intense shock waves to which they are subjected, and there is probably a corresponding reduction in fracture toughness. There is also the possibility of intermetallic alloys being formed at the weld interface, which though they may exhibit acceptable static strength might not have adequate impact strength. The thickness of the flyer and parent plates and their strength levels are such that no valid fracture toughness data can be obtained, but Charpy

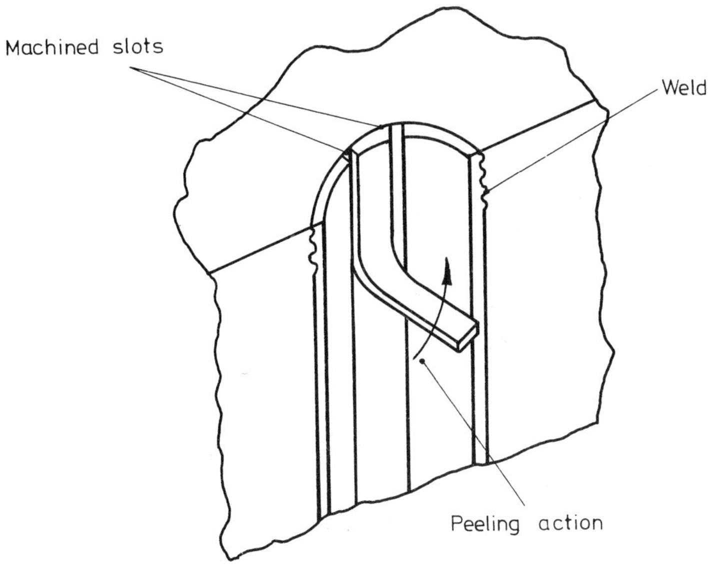

FIG. 9.9 Peel test on a tube to tubeplate weld

impact test pieces can mostly be cut from the parent plate but not the flyer plate. De Maris and Pocalyko (223) have investigated the effect of explosive welding on the impact strength of the parent plate, in which 3.2 and 4.8 mm thick 304 stainless steel was cladded to 25.4 and 28.6 mm thick A-212-B carbon steel and A-302-B and A-387-D alloy steels. As expected the shock loading reduces the Charpy impact strength of the steel parent plate, but this effect decreases as a function of the distance from the weld interface. At about 12 mm from the interface the effect of shock loading becomes negligible. A stress-relieving heat-treatment restores the impact strength of the parent plate to at least its pre-clad value. However, it is noted in Chapter VI that heat treatment will pose metallurgical problems for many metal combinations, and even with stainless steel to steel welds there may be unacceptable carbon migration, which if it is excessive will reduce the corrosion resistance.

Wylie et al (95) and Crossland and Williams (61) have used the impact specimen illustrated in Fig. 9.10 to investigate the effect of welding parameters on the strength of the bond. For this purpose it was probably an acceptable test, but the value of the impact strength measured depends on the flyer plate

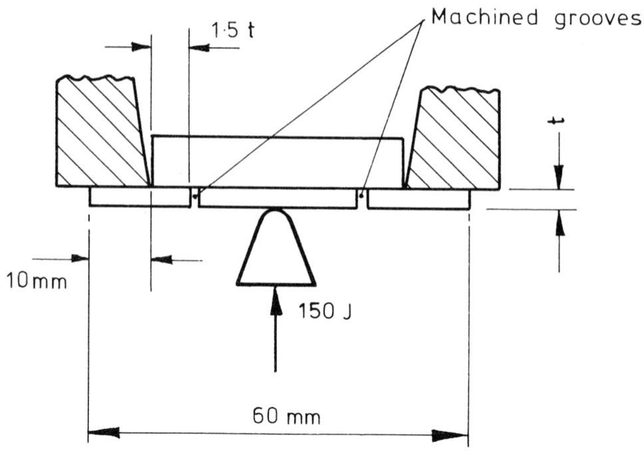

FIG. 9.10 Impact test on weld interface (95)

thickness, as a thin flyer plate will result in more plastic deformation, prior to failure and an increased peeling action. This implies that the value of impact strength established with the specimen is not solely a material property.

Some fatigue tests of explosively cladded plate have been reported by De Maris and Pocalyko (223), Gelman et al (224) and Banerjee and Crossland (225). De Maris and Pocalyko have reported on bending fatigue tests of 304-L stainless steel, Timet 35-A titanium and Hastelloy alloy all separately cladded to ASTM A-212-B FbQ steel, and Inconel 600 clad to 50 mm thick ASTM A-302-B FbQ low alloy steel. In the first three composites the specimens were machined with the weld interface as close as possible to the surface of the specimen to subject the weld interface to the maximum alternating stress, while with the last composite the weld interface was in the plane of maximum flexural shear. Specimens were tested in the as welded state and with the last composite some specimens were subjected to a stress relieving heat treatment after welding. The interpretation of the data is made difficult by the the differing elastic and thermal properties of the two materials forming the composite, and in the as welded state the presence of residual stresses of unknown magnitude. None of the fatigue fractures in these specimens appeared to initiated at the weld interface and none of the specimens showed any bond separation. It was concluded that there was no great decrease in the endurance limit attributable to the explosive welding process.

Gelman et al (224) have also carried out bending fatigue tests, but on composites consisting of steels of different mechanical properties and different coefficients of thermal expansion. The parent plate was cladded on both surfaces to give a symmetrical test piece. They found that the fatigue strength of as welded specimens was significantly reduced, but it could be raised by heat treatment after welding, and this improvement was greater for a flyer plate material which had a smaller coefficient of thermal expansion than that for the parent plate. This could be due to favourable residual stresses induced in the flyer plate material after heat treatment. No bond separation was noted.

Banerjee and Crossland (225) carried out tensile fatigue tests on stainless steel/mild steel and brass/mild steel composites in which the weld interface was on the central plane. They found with stainless steel/mild steel in the as welded state that the fatigue strength was identical to that of the stainless steel prior to cladding, which was significantly greater than for the mild steel. They also noted that the failure appeared to be initiated at the weld interface. For brass/mild steel the fatigue strength in the mortal region lay between the fatigue curves for the mild steel and brass prior to cladding, while at the fatigue limit the fatigue strength was the same as the mild steel which was the stronger of the two materials making up the composite. The authors noted that in the as welded state the fatigue properties could not be strictly compared with the properties prior to welding, as after explosive welding both materials forming the composite had been shock hardened which would affect their properties. There is also a shear stress acting at the weld interface due to differences in the moduli of the component materials.

The only work which appears to have been reported on the effect of thermal shock or thermal fatigue on the properties of explosively cladded plate is a brief feasibility study reported by Banerjee and Crossland (225). It is obviously of some importance as cladded plate is frequently used in the fabrication of heat exchangers and chemical reaction vessels. Avery (226) and Baron (227) have reviewed the subject in relation to homogeneous materials but not for composite materials, where the difference of the coefficient of expansion for the two component materials may enhance the thermal stresses created. Thermal shock occurs due to the thermal stresses arising from a sudden change of temperature, such as occurs at start-up in a gas turbine or as a result of the flash temperature when the propulsive charge in a gun initiated. Thermal shock could conceivably cause failure in a composite plate if there was a brittle intermetallic layer at the weld interface. Thermal fatigue arises from repeated cycles of heating and cooling but not necessarily under shock conditions and this might arise at the weld interface especially at the stress raisers arising from the interfacial waves or pockets of cast structure. Banerjee and Crossland subjected stainless steel/mild steel and brass/mild steel to ten cycles of heating to various temperatures followed by water quenching at 20°C. Tensile shear tests of the type shown in Fig. 9.4 were then carried out to discover if there was any deterioration in the strength of the weld. In both cases no deterioration was noted up to temperatures of 500-600°C, but above this a slight but definite reduction in strength was noted.

Destructive tests can only be carried out on specimens from cut-offs in the case of explosive cladding, but in other applications such as tube to tubeplate welding no destructive test is possible. The only well established non-destructive test is ultrasonic crack detection, but this is only capable of establishing areas of no bond, and it is not at present able to detect the presence of a poor metallurgical structure at the weld interface. Kimura and Takahashi (228) have examined the ultrasonic testing of explosive welds between steel and other metals. They noted that unless the two component metals had an acoustical impedance ratio within the range of 1.09 to 0.92 such as Hastelloy c, Monel metal, AISI 405 S/S and AISI 304 S/S, then there will be a reflection from the weld interface. For composites with an impedance ratio within this range a defective bond will be detected by the presence of a reflection from the imperfection. For a perfectly welded and defect free clad steel with an impedance ratio outside the above range, it was noted that the ratio of amplitudes of the reflection from the weld interface to that from the back surface could be expressed as a function of the acoustical impedances of the two component metals. When the weld is defective then the reflection from the weld interface will be larger and that from the back surface smaller, so the ratio of amplitudes will increase and this increase will be a measure of the imperfection.

Ultrasonic testing using a hand held miniature probe is commonly applied to tube to tubeplate welds. Robinson et al (229) have developed an automated system for the examination of tube to tubeplate welds. The detection of front crevices in tube to tubeplate or plug to tubeplate welds is of importance in stainless steel systems in the presence of wet steam, as such crevices can initiate crevice corrosion. Ultrasonic inspection is not adequate for the detection of small crevices, and the only valid inspection technique is a dye penetrant test.

The mechanical testing of the strength of explosive welds has been widely reported and, in general, if the welding is carried out with the correct parameters then the strength is more than adequate for its duty. Under correct welding conditions the static strength is equal to the weaker of the two component materials, which is the most that can be expected. Without heat treatment subsequent to the welding operation the impact strength and probably the fracture toughness as well will be reduced by the shock loading to which the materials are subjected. If the properties of the component material are sufficiently high, then the drop of impact strength as a result of welding may be tolerable without the need for a subsequent heat treatment. However, the one factor which may affect the strength and the corrosion resistance of explosively cladded plate is the residual stresses induced in the cladded plate by the welding process.

Residual stresses in explosive welds have been investigated by Taksukawa (230), Taksukawa and Oda (231), Pruemmer (232) and Pathiraj and Vasudevan (233). Taksukawa examined the residual stresses in a mild steel/mild steel weld, and he observed residual tensile stresses throughout the thickness of the flyer plate and corresponding residual compressive stresses in the parent plate. Taksukawa and Oda observed somewhat similar residual stresses in a stainless steel/mild steel weld except that close to the surface of the stainless steel flyer plate the residual stress was compressive. Perhaps more importantly they measured the residual stresses after annealing at 610°C for one hour and found that they were very little different except that those in the stainless steel flyer plate were wholly tensile. They attributed this effect to the relative thermal expansion at 610°C causing plastic distortion. It clearly demonstrates that for this particular case there is nothing which can be done to remove the residual stresses caused by welding. Pruemmer looked at the residual stresses induced in welding a mild steel flyer plate to aluminium alloy and within 4 mm of the weld interface he found that the stresses in both the parent and flyer plates were compressive, but these must be balanced by tensile stresses at greater distances from the weld interface. Pathiraj and Vasudevan examined the residual stresses induced in welding nearly pure aluminium to steel, and they observed residual tensile stresses in the steel and residual compressive stresses in the aluminium. In all the cases the magnitude of the maximum residual stress lay between 200 and 400 MPa which is of the magnitude of the yield stress of the materials concerned. Stresses of this magnitude can hardly be ignored in the assessment of the strength and corrosion resistance of explosively fabricated composites.

The particular advantage of explosive welding as with other solid phase welding processes is the ability to join otherwise incompatible alloys, which because of differences in the melting temperatures and the formation of harmful intermetallic alloys could not be fusion welded. However, because the more severe metallurgical problems associated with fusion welding are avoided by explosive welding, does not imply that there are no metallurgical problems, but that the problems are different.

There is a considerable literature relating to the metallurgy of explosive welding and associated fields, which is reviewed by Crossland and Williams (17). Additionally a discussion of various aspects of the problems are given in papers by Lucas and Williams (234), Lucas, Williams and Crossland (89), Williams, Dhir and Crossland (235) and Dhir,

Williams and Crossland (236). However, the subject is too important to be ignored in a book on explosive welding, and it will now be briefly reviewed.

In explosive welding the flyer plate is subjected to an intense stress wave from the detonation of the explosive in contact with it, and subsequently both the flyer plate and parent plate experience an intense stress wave resulting from the high velocity impact. These intense stress waves give rise to metallurgical changes which result in an increase in hardness, which is illustrated in Fig. 9.11. The general level of hardness in both plates is increased and if high detonation velocity explosive is used which gives a much more intense pressure, then the surface of the flyer plate in contact with the explosive will show a higher hardness unless an adequate buffer is used. Fig. 9.11a shows a general increase in hardness with a localised hardening at the interface duc, maybe, to the intense shock pressure on collision or the intense plastic deformation in the region of the weld interface. Alternatively Fig. 9.11b shows another profile in which the hardness close to the interface is less than the general level in the flyer and

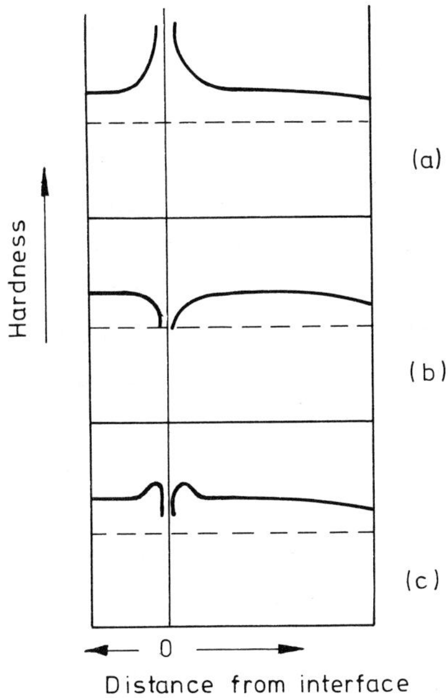

FIG. 9.11 Typical micro-hardness profiles in welds between similar metals. ---- hardness before welding

parent plate, due to the heat generation in the region of the interface which causes annealing of the work hardened material. The last possibility is shown in Fig. 9.11c which is an intermediate case between the two extremes, in which partial annealing of the work hardened weld zone has occurred. Examples of all three hardness profiles shown in Fig. 9.11 have been reported in numerous papers, for instance Pearson (10), Pearson and Hayes (11) and Bahrani and Crossland (220).

Besides an increase in hardness in the flyer and parent plates, which is associated with an increase in tensile strength, there is a reduction in ductility and impact strength. De Maris and Pocalyko (223) and Loosemore and Peapell (237) have investigated impact strength of the steel parent plate in explosive welds, and they have observed a reduction in the lower and upper shelf impact energies, and an increase in the ductile/brittle transition temperature. Loosemore and Peapell noted that this reduction of impact properties increased with explosive loading, so if it is desired to try and avoid heat-treatment after welding then it is obviously desirable to use the minimum explosive loading which provides an acceptable weld.

The problem is that these increases in hardness and tensile strength and the corresponding reduction in ductility and impact strength occur with little gross deformation and no visible grain distortion. Obviously these changes must be associated with the passage of the intense compressive wave to which the materials are subjected or the reflected rarefaction wave. There is a considerable literature relating to shock waves in solids, but much of this is concerned with much higher pressures than are experienced in the intense stress waves generated in explosive welding. In explosive welding the pressure associated with the intense stress waves rarely exceeds 20 GPa, and under normal welding conditions it will be appreciably less.

Smith (238), Smith and Fowler (239), Zukas and Fowler (240), Tardif et al (241), Dieter (242), Holtzman and Cowan (243), Appleton et al (244), Curran et al (245), Cohen et al (246), Bowden and Kelly (247), Brillhart et al (248) and Otto and Mikesell (249), to mention but a few have reported on the effects of shock loading on solids. At the pressures associated with explosive welding mechanical twinning can be observed in copper at pressures as low as 2.8 GPa according to Johari and Thomas (250), and also in pure iron and steels at quite low pressures. In both cases the amount of twinning is reduced by prior cold work or other hardening processes, according to Dieter (242). Fig. 9.12 shows a typical example

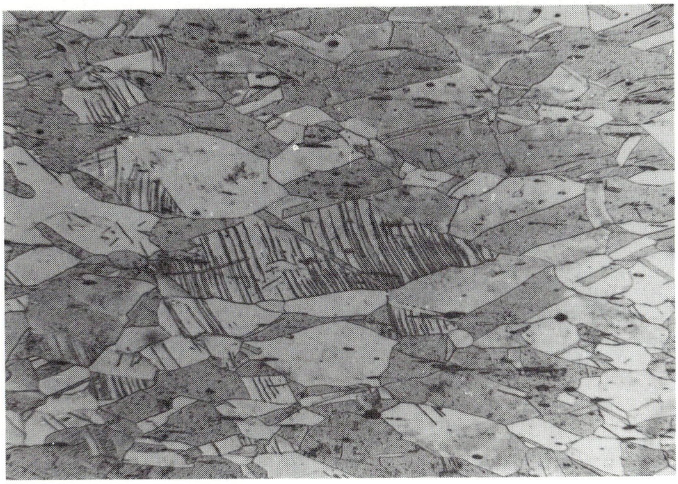

FIG. 9.12 Macro twinning in copper/copper weld (265) x 250

of mechanical twinning in the copper parent plate of a copper to copper weld. Twinning in pure iron and in carbon and alloy steel can also be induced by the reversible phase change at about 13.3 GPa in pure iron. At this pressure body centred cubic α-ferrite transforms to hexagonal close-packed ε-martensite, and this causes the twinning which is shown in Fig. 9.13. Shock hardening of austenitic manganese steel gives a remarkable increase in hardness and tensile properties, as reported by Otto and Mikesell (249). This is used commercially as a hardening process for hardening components made from this material. For this material it has been suggested from examination of the microstructural markings there is the possibility that to a limited extent there is a transformation from face centre cubic to close-packed hexagonal martensitic structure.

Metals with widely different properties and characteristics, such as iron, carbon steels, alloy steels, maraging steel, copper, copper based alloys and aluminium, show varying degrees of increase in hardness and tensile strength and reduction in

FIG. 9.13 Mechanical twinning in iron/iron weld due to α-ε-α phase change (265) x 100

ductility and impact strength when subjected to shock loading. These changes in properties occur with little gross deformation and no visible grain distortion, so the change in properties cannot be related to cold working. It is tempting to associate the increase in hardness with twinning, but Smith (238) noted in iron that there was no difference in micro-hardness between heavily and lightly marked grains, and aluminium exhibits shock hardening even though it does not develop twins. Yet another possibility is that the pressure-induced phase change, such as in iron, causes a high degree of lattice strain which will produce a high hardness. However, this would only explain the step increase in hardness experienced in the vicinity of the transition pressure, and it would not explain the hardening noted in materials without a phase transition. Lastly, theory predicts that fast moving dislocations are efficient producers of vacancies and interstitials, see Weertman (251). The presence of a high residual concentration of point defects may be expected to contribute to an increase in hardness and a decrease in ductility. This seems the most likely explanation.

The intense compression wave is reflected from a free surface as a rarefaction wave, which if it is sufficiently strong and the material sufficiently weak in tension will cause a spalling or scabbing failure, in which a scab becomes detached from the back side of the plate. This is the basis of a well known form of attack on armour plate, which consists of placing an explosive charge in contact with the outer surface of the armour plate, which when it is detonated causes a large scab to detach itself from the inner surface and cause severe havoc inside the tank. The subject of fracture of solids under impulsive loading is discussed by O'Brien and Davis (252). In explosive welding the intensity of the compressive stress wave is not normally large enough to give rise to a sufficiently intense reflected tensile wave to cause spalling. The only exception is when the flyer plate is accelerated by a high detonation velocity explosive such as a PETN sheet explosive, which in the absence of an adequate buffer induces a very intense compression wave in the flyer plate. This intense compression wave is, while the flyer plate is still in flight reflected as a strong tensile wave, which can cause a spalling failure such as that shown in Fig. 9.14. In this case the scab did not become detached. However, normally in explosive welding the compressive stress wave is not adequately strong to cause scabbing failures.

FIG. 9.14 Spalling failure in copper flyer plate x 100

Under certain conditions the weld interface will be straight and of high strength. A typical example is shown in Fig. 9.15. In this example the jet has probably escaped and

FIG. 9.15 Copper/copper weld with straight interface (265) x 100

the structure close to the weld interface consists of highly elongated grains. If the jet generated in the high velocity impact does not completely escape, then it is partially or completely trapped in the vortices formed before and after each wave in a weld between metals of similar density, or in the single vortex before or after each wave in a weld between metals of appreciably different densities. Fig. 9.16 is a

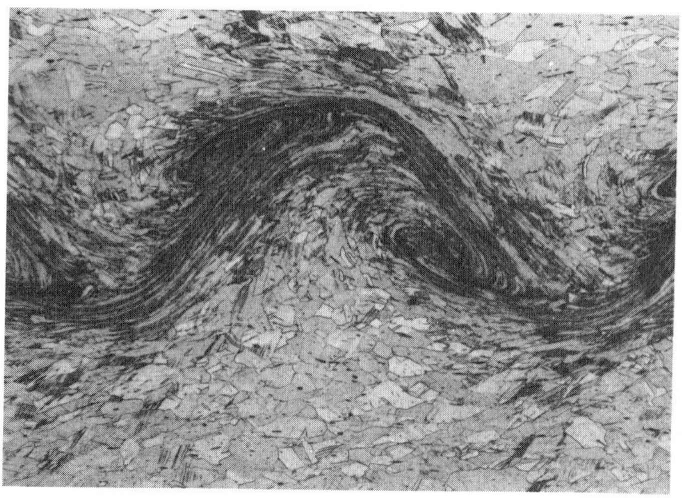

FIG. 9.16 Copper/copper weld with normal wavy interface (265) x 150

typical example of a weld between metals of similar density. The very high kinetic energy in the jet will be dissipated as heat in the vortices, and this causes melting in the centre of the vortices, and even though it is not obvious there is probably some melting in the example shown in Fig. 9.16. If the kinetic energy is greater then the melting at the centre of the vortices is more obvious, as in Fig. 9.17, and a cast

FIG. 9.17 Copper/copper weld with molten pockets in the centre of the vortices (265) x 100

structure with columnar crystal growth and cooling cavities can be observed. If the kinetic energy is further increased these molten zones will coalesce to form a continuous cast interlayer, which is shown in Fig. 9.18. Such an interface will show all the weaknesses associated with cast structures, particularly in dissimilar metal welds where harmful intermetallics can be formed. It is also possible, maybe due to jet interaction with the surface roughness, for a very thin molten layer to be formed along the weld interface.

In the vortices the molten mixture is vigorously stirred, and as it is a molten zone surrounded by relatively cold metal it is subjected to a very high cooling rate which may be comparable or even more severe than in splat cooling. Under these conditions it is to be expected that metastable compounds can be formed in these zones.

The nature of the structure in the interfacial zone and of

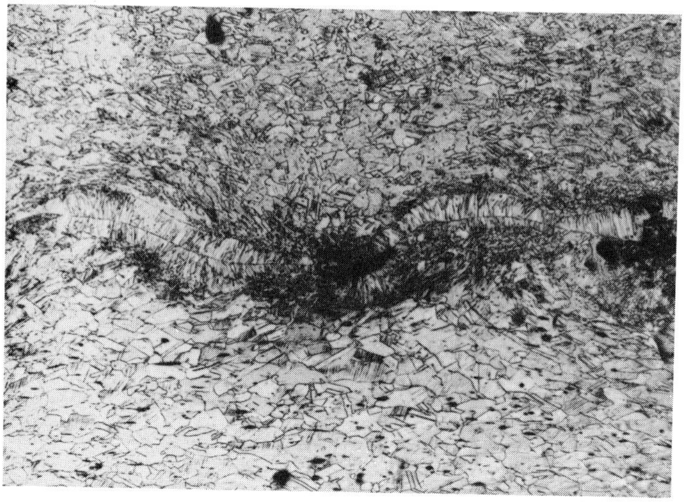

FIG. 9.18 Copper/copper weld with continuous cast interlayer (265) x 100

the bond in explosively welded composites has been the subject of extensive researches by, for instance, Hayes and Pearson (253), Davenport (254), Boes (255), Wright and Bayce (192), Van Wely and Verbraak (256), Carlson (257), Holtzman and Cowan (194), Deribas et al (258), Buck and Hornbogen (259) (260), Crossland and Bahrani (261), Pietteur (262), Lucas and Williams (233), Berdichevskii and Sobolenko (263), Trueb (264), Lucas et al (234), Williams et al (235), Dhir et al (236). However, there appears to be considerable differences in the results obtained and the conclusions reached even for welds between nominally identical pairs of metals.

Looking at the weld interfaces in Figs. 9.15 - 9.18 for a simple copper/copper weld it should occasion no surprise that different workers have reached different conclusions, as the nature of the interface depends on the particular point on a particular weld which is selected for examination. The nature of the weld interface will depend upon the explosive loading, the properties of the explosive used, the experimental configuration, the thickness of the flyer plate and the properties of the flyer and parent plates. If metals of very different densities are being welded then there is a greater probability of a thin cast layer along the whole interface, which may enhance diffusion or which may itself be mistaken for a zone of diffusion. If it is a very thin interlayer then an electron beam analysis might be misleading if the beam diameter is comparable to the thicknesses of the interlayer, and it has to be recognised that it cannot differentiate between an intermetallic compound and a finely divided mixture of particles of the two component materials. X ray analysis of the thin layer is very difficult or maybe impossible unless the thin layer can be separated from the rest of the material, but it is more suitable for the analysis of the materials to be found in the vortex zones.

Fig. 9.19b from Lucas et al (234) shows an electron micro-

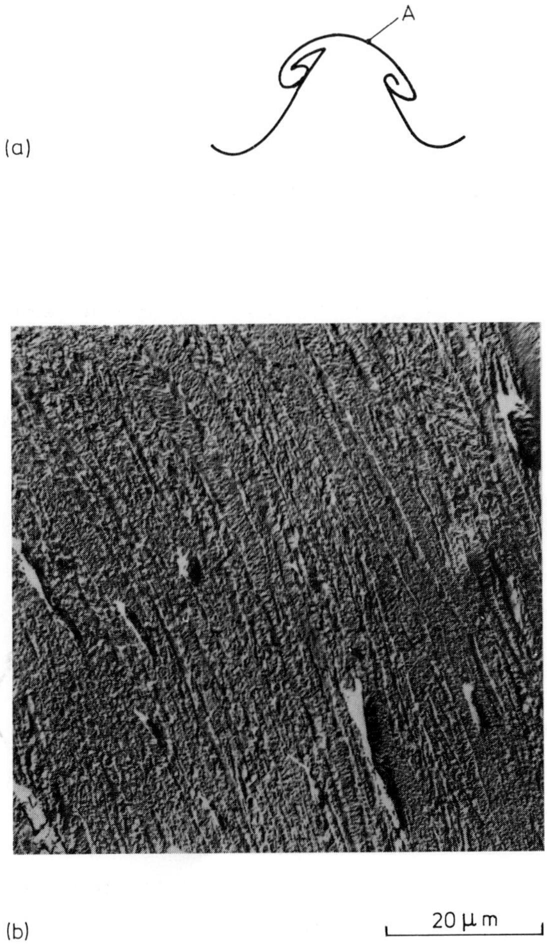

(c)

FIG. 9.19 Copper/copper weld (33). (a) waveform, (b) electron micrograph at interface at A, (c) thin film transmission electron micrograph at A

graph of a copper to copper weld at a point on the interface between the front and rear vortices. It will be noted that the bond zone consists of severely elongated grains. Based on the reduction of thickness of the grains the approximate strain at 400%. Heat treatment at 250°C for ½ hour resulted in complete recrystallisation of the bond zone but no recrystallisation was observed in the shocked and vortex zones, from which it was deduced that the bond zone in the as welded conditions was intensely deformed and largely unrecrystallised. A transmission micrograph of the same region is shown in Fig. 9.19c, which gives evidence of some recovery and recrystallisation.

In the region of the vortices the cast structure is made up of columnar crystals, sometimes with a central cooling cavity. On the periphery of the central cast zone the structure rapidly changes to become similar to the severely strained structure observed at the interface remote from the vortices as shown in Fig. 9.19b, except that the strains are not so large.

In welds between dissimilar metals a variety of bond zones

can be observed depending on the particular combination. Figs. 9.20 and 9.21 show the interface away from the vortices for

FIG. 9.20 Electron micrograph of copper/nickel weld

FIG. 9.21 Electron micrograph of stainless steel/mild steel weld (33)

copper/nickel and stainless steel/mild steel welds, which because the component materials have nearly equal densities exhibit a front and rear vortex. In both cases the joint between the two metals resembles a grain boundary. An electron probe analysis across such a bond shows no detectable diffusion if due allowance is made for the diameter of the electron beam. In the stainless steel/mild steel weld the structure of the pearlite grains in the mild steel show no signs of breaking down due to carbon diffusion into the stainless steel or chromium/nickel diffusing into the mild steel, which indicates that if there is diffusion it is insignificant.

Fig. 9.22 shows the interface in an aluminium/copper weld

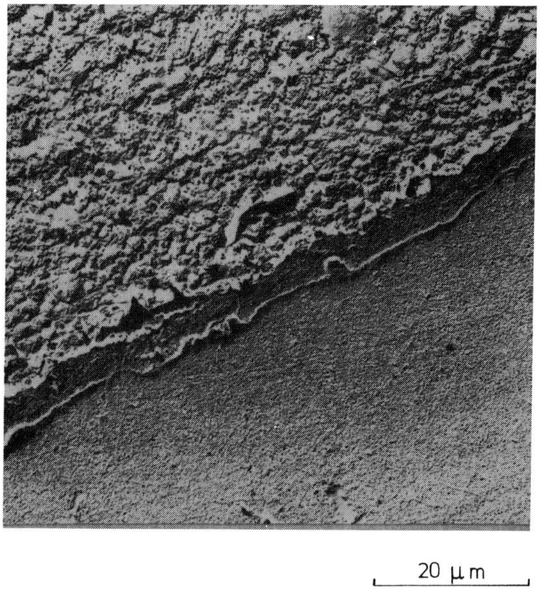

FIG. 9.22 Electron micrograph of aluminium/copper weld (33)

which, as the component materials are of very different densities, exhibits a cast layer along the interface. It was verified that this was not due to a dissimilar etching zone by examining a corresponding cold pressure weld, which exhibited no layer at the interface. An electron probe analysis of both these specimens clearly demonstrated that there was a cast layer in the explosively welded specimen, in which the copper and aluminium content varied smoothly from 100% of one component and 0% of the other to 0% and 100% across the layer.

There were no peaks in these transition curves, which suggests that there is no intermetallic compounds. Lucas (33) also examined another sample where large scale melting had occurred and there were four peaks in the electron probe analysis traverse of the cast zone. A low temperature anneal of one hour at 150°C was carried out and this made no difference to the electron probe analysis, so it was concluded that there was no metastable phases. Two of the peaks noted appeared to be realistic compounds but the other two could not be accounted for, and probably they were due to mechanical mixture of the two components.

Fig. 9.23a shows the interface of a titanium/mild steel explosive weld which exhibits a continuous melted interlayer. Williams (265) noted when a titanium specimen was tested in shear that some of the interlayer material became detached. He subjected samples of this material to X-ray analysis and found that the diffraction pattern corresponded closely with the hexagonal close-packed compound $TiFe_2$, which was in agreement with earlier work, reported by Deribas et al (258). This compound is very hard; Crossland and Bahrani (261) for instance recorded a hardness of 1290 Hm, and it is also unacceptably brittle as can be seen in Fig. 9.23a where cracks can be seen which were initiated from the micro-hardness indentations. Such a weld would be unacceptable, but it is possible to produce a titanium/mild steel weld such as that shown in Fig. 9.23b in which there is only a small zone of intermetallic compound wrapped up in the centre of the vortex.

The vortices at the interface in explosive welds contain a mixture of both component materials, and from the wave mechanism proposed by Bahrani et al (29) it would be expected that the rear vortices would mainly contain the surface of the flyer plate while the front vortices would contain the parent plate surface. This was confirmed by Crossland and Bahrani (261) who carried out electron probe analysis of the material in the front and rear vortices of a stainless steel/mild steel weld, and noted that the percentage of stainless steel was 70% in the rear vortex and 30% in the front vortex. Lucas et al (234) carried out tests in which the mild steel flyer plate was plated with a thin film of nickel and the mild steel parent plate was

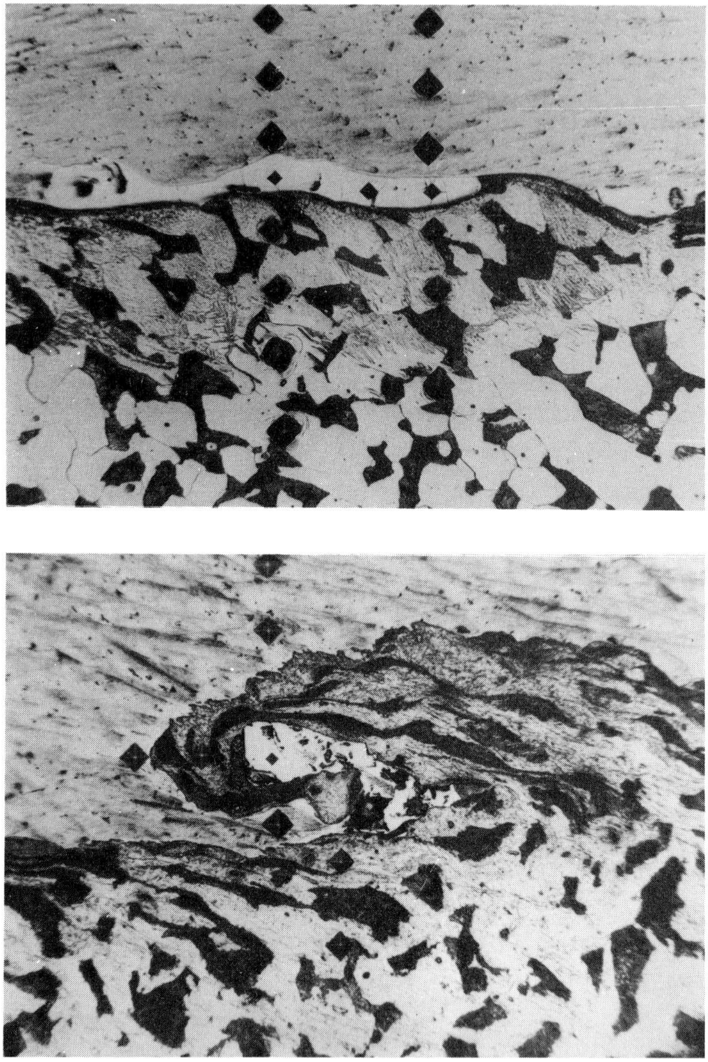

FIG. 9.23 Interface of a titanium/mild steel weld showing the hard intermetallic compound (216). (a) continuous layer at interface x 200, (b) front vortex x 150

plated with copper, and this again confirmed the deductions from the wave mechanism of Bahrani et al. Micro-probe analysis carried out by Lucas (33) suggested that the ratio of flyer plate to the parent plate is generally 6:4 in the rear vortex and 3:7 in the front vortex for welds between materials of similar density. Teslenko et al (126) examined ten melted zones in a nickel to steel weld and found that no two had exactly the same composition. They noted that when the vortex entered the steel the concentration of steel was 60 - 75% and where the vortex entered the nickel the concentration was marginally reduced to 50 - 60%. However, Williams et al (235) while agreeing in general with Lucas found that the precise composition in the vortices can vary fairly widely depending on welding conditions.

Williams et al (235), Dhir et al (236) and Dhir (266) have made theoretical estimates of the cooling rates in the molten zones in the vortices, and they have concluded that they are of the order 10^5-10^7 $^{o}C/s$ during and immediately after solidification. These cooling rates are similar to those estimated for splat cooling conditions where extreme deviations from equilibrium structures are common. They carried out X-ray diffraction on samples from the vortex zones which were removed from the specimen by grinding and chemically dissolving one of the component metals. They found metastable phases in silver/copper, silver/iron, iron/copper, iron/lead, cadmium/aluminium, tin/aluminium, cadmium/lead and silver/lead explosive welds. In all cases they found that these metastable phases appeared to be stable at room conditions, but they all decomposed after heating the specimen to an appropriate temperature. In the cases where there was a single phase present the analysis was determined by microprobe analysis, but when there was a duplex structure the microprobe analyser could not discriminate between the phases so no analysis was possible.

From this brief review of the metallurgy of explosive welds it appears that the diffusion across the bond during the process is in general negligible. In welds between metals of very different densities there is a greater possibility of the formation of what appears to be a thin cast interlayer, which may consist of intermetallic compounds with unfavourable mechanical properties. The vortex zones contain a mixture of the two component metals and due to the kinetic energy of the trapped jet there is frequently a molten zone at the centre, which frequently contains metastable intermetallic compounds. However, though the mechanical properties of these compounds may be poor the effect on the overall properties will be small if the compound is wrapped up in the vortex, and if it is surrounded by material with acceptable mechanical properties. As has been seen in Chapter VI if heat treatment is required to remove the ill effects of shock loading or after cold working or if the service temperature is high, then diffusion can occur across the interface and indeed it will be enhanced by the plastic deformation and defect density and structure at the interface.

BIBLIOGRAPHY

1. Singer, C., Holmyard, E.J. and Hall, A.R. (1954), A History of Technology, Oxford University Press.
2. Houldcroft, P.T. (1977), Welding Process Technology, Cambridge University Press.
3. Tylecote, R.F. (1968), Solid Phase Welding of Metals, Edward Arnold.
4. Carl, L.R. (1944), Brass Welds, Made by Detonation Impulse, Metal Progress, 46, 102-103.
5. Deribas, A.A. (1967), Explosive Welding, Siberian Branch of Academy of Sciences, U.S.S.R.
6. Phillipchuk, V. (1965), Explosive Welding Status, 1965 ASTME Creative Manufacturing Seminar, paper SP65-100.
7. Phillipchuk, V. and Le Roy Bois, F. (1962), Explosive Welding, U.S. Patent 3 024 526.
8. La Rocca, E.W. and Pearson, J. (1958), Explosive Press for use in Impulsive Loading Studies, Rev. Sci. Instr., 29, 848-851.
9. Pearson, J. (1960), Metal Working with Explosives, J. Metals, 12, 673-681.
10. Pearson, J. (1961), Explosive Welding, 1961 Advanced High Energy Rate Forming, Book I, paper SP60-159.
11. Pearson, J. and Hayes, G.A. (1963), Research in Explosive Welding, 1963 ASTME National Welding and Metals Joining Conference, paper SP63-97.
12. Allen, W.A., Mapes, J.M. and Wilson, W.G. (1954), An Effect Produced by Oblique Impact of a Cylinder on a Target, J. Appl. Phys., 25, 675-676.
13. Davenport, D.E. and Duvall, G.E. (1961), 1961 Advanced High Energy Rate Forming, Book I, paper SP60-61.
14. Cowan, G.R., Douglass, J.J. and Holtzman, A.H. (1962), U.S. Patent 3 137 937.
15. Tardif, H.P. (1960), Explosive Forming in Canada, Metal Progress, 77, No. 1, 128-130.
16. Crossland, B. and Bahrani, A.S. (1968), Explosives and Their Use in Engineering, Metals and Materials, 2, 41-49

and 68-79.

17. Crossland, B. and Williams, J.D. (1970), Explosive Welding, Metallurgical Reviews, 15, 79-100.

18. Crossland, B. (1971), The Development of Explosive Welding and its Application, Metals and Materials, 5, 401-413.

19. Procs. of the 1st, 2nd, 3rd, 4th, 5th, 6th and 7th Int. Conf. of the Centre for High Energy Forming (1967, 1969, 1971, 1973, 1975, 1977, 1981), Denver Research Institute.

20. Procs. of the Meetings of the High Pressure Technology Association (1968, 1973), Use of Explosives in Forming Welding and Compaction, Belfast, and Int. Conf. on the Use of High-Energy Rate Methods for Forming, Welding and Compaction, Leeds.

21. Procs. of Conference of the Welding Institute (1969, 1976), Select Conference on Explosive Welding, Hove, and Conference on Explosive Welding, London.

22. Procs. of the 1st, 2nd and 3rd Int. Symposium on Use of Explosive Energy in Manufacturing Metallic Materials of New Properties and Possibility of Application Thereof in Chemical Industry, Marianske Lazne, Czechoslovakia. (1970, 1973, 1976).

23. Birkhoff, G., MacDougall, D.P., Pugh, E.M. and Taylor, G. (1948), Explosives with Lined Cavities, J. Appl. Phys., 19, 563-582.

24. Cowan, G.R. and Holtzman, A.H. (1963), Flow Configuration in Colliding Plates: Explosive Bonding, J. Appl. Phys., 34, 928-939.

25. Bergmann, O.R., Cowan, G.R. and Holtzman, A.H. (1966), Experimental Evidence of Jet Formation During Explosion Cladding, Trans. Met. Soc. of AIME, 236, 646-653.

26. McKee, F.A. and Crossland, B. (1975), Further Experiments on the Mechanism of Explosive Welding, Proc. 5th Int. Conf. of High Energy Rate Fabrication, 4.11.1-4.11.25.

27. Walsh, J.M., Shreffler, R.G. and Willig, F.J. (1953), Limiting Conditions for Jet Formation in High Velocity Collision, J. Appl. Phys., 24, 349-359.

28. Abrahamson, G.R. (1961), Permanent Periodic Surface Deformations due to a Travelling Jet, J. Appl. Mech., 83,

519-528.

29. Bahrani, A.S., Black, T.J. and Crossland, B. (1967), The Mechanics of Wave Formation in Explosive Welding, Proc. Roy. Soc. (A), 296, 123-136.

30. Williams, J.D., Private Communication.

31. Wilson, M.P.W. and Brunton, J.H. (1970), Wave Formation between Impacting Liquids in Explosive Welding and Erosion, Nature, 226, 538-541.

32. Kudinov, V.M. and Bunyatyan, A.Kh. (1973), Hydrodynamic Modelling of Wave Formation at the Interface of Explosive Welds, Int. Conf. on the Use of High-Energy Rate Methods for Forming, Welding and Compaction, University of Leeds, 8.1-8.2.

33. Lucas, W. (1970), A Metallurgical Investigation of Explosion Welding, Ph.D. Thesis of the Queen's University of Belfast.

34. Homann, Von F. (1936), Einfluß großer Zühigkeit bei Stromung um Zylinder, Forschung auf dem Gebiete des Ingenieurwesens, 7, 1-10.

35. Cowan, G.R., Bergman, O.R. and Holtzman, A.H. (1971), Mechanism of Bond Zone Wave Formation in Explosion-Clad Metals, Met. Trans., 2, 3145-3155.

36. Kowalick, J.F. and Hay, D.R. (1971), A Mechanism of Explosive Bonding, Met. Trans., 2, 1953-1958.

37. Reid, S.R. (1974), A Discussion of the Mechanism of Interface Wave Generation in Explosive Welding, Int. J. Mech. Sci., 16, 399-413.

38. Reid, S.R. and Sherif, N.H.S. (1976), Prediction of the Wavelength of Interface Waves in Symmetric Explosive Welding, J. Mech. Eng. Sci., 18, 87-94.

39. Bahrani, A.S. and Crossland, B. (1964), Explosive Welding and Cladding: An Introductory Survey and Preliminary Results, Inst. Mech. Engr., 179, 264-281.

40. Harlow, F.H. and Pracht, W.E. (1966), Formation and Penetration of High-Speed Collapse Jets, Phys. Fluids, 9, 1951-1959.

41. Klein, W. (1970), The Flow Processes in the Contact Zone during Explosion Cladding of Metals, Doctorate Thesis of the Technical University, Clausthal.

42. Hunt, J.N. (1968), Wave Formation in Explosive Welding, Phil. Mag., 17, 669-680.

43. Onzawa, T. and Ishii, Y. (1973), Study of the Wave Formation in Explosive Bonding, Trans. J. W. S., 4, 233-240.

44. Onzawa, T. and Ishii, Y. (1975), Fundamental Studies on Explosive Welding - Observations of Metal Jet and Wavy Pattern, Trans. J. W. S., 6, 98-104.

45. Klein, W. (1971), The Flow Path in the Contact Zone in Explosion Welding, Fried-Krupp G.M.B.H.

46. Keller, K. (1967), Contributions to Explosion Cladding, Dissertation Technische Hochscule, Stuttgart.

47. Godunov, S.K., Deribas, A.A., Zabrodin, A.V. and Kozin, N.S. (1970), Hydrodynamic Effects in Colliding Solids, Computational Physics, 5, 517-539.

48. Deribas, A.A. (1972), Some Phenomena Accompanying High-Speed Collision of Solids, Proc. 13th Int. Cong. of Theoretical and Appl. Mech., Moscow, 72-84.

49. El-Sobky, H. and Blazynski, T.Z. (1975), Experimental Investigation of the Mechanics of Explosive Welding by Means of a Liquid Analogue, Proc. 5th Int. Conf. on High Energy Rate Fabrication, 4.5.1-4.5.21.

50. Chadwick, M.D., Howd, D., Wildsmith, G. and Cairns, J.H. (1968), Explosive Welding of Tubes and Tubeplates, Brit. Weld. Jl., 15, 480-492.

51. Taylor, J. (1952), Detonation in Condensed Explosives, Oxford University Press.

52. Cook, M.A. (1958), The Science of Explosives, Reinhold.

53. Rinehart, J.S. and Pearson, J. (1963), Explosive Working of Metals, Pergamon Press.

54. Fordham, S. (1980), High Explosives and Propellants, Pergamon Press.

55. Johansson, C.H. and Persson, P.A. (1970), Detonics of High Explosives, Academic Press.

56. Dautriche, H. (1906), Sur les Vitesses de Detonation des Explosifs, Compt. Rend., 143, 641-644.

57. Minshall, S. (1955), Properties of Elastic and Plastic Waves Determined by the Pin Contactors and Crystals, J. Appl. Phys., 26, 463.

58. Marshall, A. (1917), Explosives, J. and A. Churchill
59. Taylor, J. and Gay, P.F. (1958), British Coal Mining Explosives, Newnes.
60. Shribman, V. and Crossland, B. (1969), An Experimental Investigation of the Velocity of the Flyer Plate in Explosive Welding, Proc. 2nd Int. Conf. of the Center for High Energy Forming, 7.3.1-7.3.37.
61. Crossland, B. and Williams, P.E.G. (1973), Explosive Welding of Tubes to Tubeplates, Proc. 2nd Int. Conf. on Pressure Vessel Technology, 1131-1149.
62. Handbook on Anfo (1967), The New Look of Blasting with Ammonium Nitrate - Fuel Oil Mixtures, Gulf Oil Corporation.
63. Anfo Manual (1972), Anfo, its Explosive Properties and Field Performance Characteristics, Monsanto Company.
64. Crossland, B. and Cave, J.A. (1975), The Explosive Properties of Ammonium Nitrate/Fuel Oil Mixtures, Proc. 5th Int. Conf. on High Energy Rate Fabrication, 4.9.1-4.9.11.
65. An Act to Amend the Law with Respect to Manufacturing, Keeping, Selling, Carrying and Importing Gunpowder, Nitro-glycerine and Other Explosive Substances (1875), The Law Reports, 10, 144-215.
66. An Act to Amend the Explosive Act of 1875 (1923), The Law Reports, 16, 277-278.
67. Veil, S.H. (1960-61), Safety and High Energy Rates, ASTME Creative Manufacturing Seminars, paper SP60-103.
68. Thomas, H. (1961), Safety with Explosive Forming, Metalworking Production, 105, 51-52.
69. Ezra, A.A. (1973), Principles and Practice of Explosive Metalworking, Industrial Newspapers Ltd.
70. Yamamoto, K. (1974), Silencer Structure for Use with Explosives, Brit. Patent 1 376 763.
71. Kolsky, H. (1953), Stress Waves in Solids, Oxford University Press, also published 1963 by Dover.
72. Rinehart, J.S. and Pearson, J. (1954), Behaviour of Metals under Impulsive Loads, American Society of Metals, also published 1964 by Dover.
73. Goldsmith, W. (1960), Impact, Edward Arnold.
74. Shewman, P.G. and Zackay, V.F. (1961), Proc. of Conference

on Response of Metals to High Velocity Deformation, Interscience.

75. Johnson, W. (1972), Impact Strength of Materials, Edward Arnold.

76. Wasley, R.J. (1973), Stress Wave Propagation in Solids, Marcel Dekker.

77. Rinehart, J.S. (1975), Stress Transients in Solids, Hyperdynamics.

78. Kaye, G.W.C. and Laby, T.H. (1973), Tables of Physical and Chemical Constants and Some Mathematical Functions, Longmans.

79. Walsh, J.M., Rice, M.H., McQueen, R.G. and Yarger, F.L. (1957), Shock Compression of Twenty Seven Metals, Phys. Rev., 108, 196-216.

80. Evans, W.M. (1970), Private Communication.

81. Duvall, G.E. and Fowles, G.R. (1963), Shock Waves, High Pressure Physics and Chemistry, edited by Bradley, R.S., 2, 209-291, Academic Press, New York.

82. Jones, O.E. (1972), Metal Response under Explosive Loading, Proc. of 12th Annual Symposium of the New Mexico Section of ASME on Behaviour and Utilization of Explosives in Eng. Design, 125-148.

83. Gurney, R.W. (1943), The Initial Velocities of Fragments from Bombs, Shells and Grenades, BRL Report 405.

84. Kennedy, J.E. (1972), Explosive Output for Driving Metal, Proc. of 12th Annual Symposium of the New Mexico Section of ASME on Behaviour and Utilization of Explosives in Eng. Design, 109-124.

85. Takizawa, Yu, Izuma, T., Onzawa, T. and Fujita, M. (1975), An Experimental Study of the Acceleration Zone and the Terminal Velocity of Flyer Plate Driven by Explosives, Proc. 5th Int. Conf. on High Energy Rate Fabrication, 4.18.1-4.18.42.

86. Smith, E.G. and Linse, V.D. (1977), The Acceleration Characteristics of Explosively Driven Flyer Plates, presented at the 6th Int. Conf. on High Energy Rate Fabrication.

87. Shorshorov, M.Kh. and Krasulin, YU.L. (1968), Calculations of Pressure Welding Parameters, Annual Meeting of the Int.

Inst. of Weld., Commission I document.

88. Shribman, V., Bahrani, A.S. and Crossland, B. (1969), The Techniques and Mechanism of Explosive Welding, The Prod. Engr., 48, 69-83.

89. Lucas, W., Williams, J.D. and Crossland, B. (1969), Some Metallurgical Observation on Explosive Welding, Proc. 2nd Int. Conf. of the Centre for High Energy Forming, 8.1.2-8.1.37.

90. Sedikh, V.S. and Bondar, M.P. (1963), Basic Parameters of Explosive Welding and Strength Characteristics of Welded Joints, Weld. Prod., 10, 1-5.

91. Carpenter, S., Wittman, R.H. and Carlson, R.J. (1967), The Relationship of Explosive Welding Parameters to Material Properties and Geometry Factors, Proc. 1st Int. Conf. of the Centre for High Energy Forming, 1.2.1-1.2.23.

92. Kowalick, J.F. and Hay, D.R. (1969), Explosive Bonding: Dimensional Analysis, Proc. 2nd Int. Conf. of the Centre for High Energy Forming, 7.2.1-7.2.18.

93. Kowalick, J.F. and Hay, D.R. (1969), Metallographic Measurement of Explosive Welding Parameters, Proc. Select Conf. on Explosive Welding, pub. Welding Institute, 78-82.

94. Crossland, B. and Williams, J.D. (1970), Welding Parameters for Explosive Cladding, Proc. Conf. Advances in Welding Processes, pub. Welding Institute, 78-82.

95. Wylie, H.K., Williams, P.E.G. and Crossland, B. (1970), (1971), Further Experimental Investigation of Explosive Welding Parameters, Proc. 1st Int. Sym. on Use of Explosive Energy in Manufacturing Metallic Materials of New Properties and Possibilities of Application thereof in the Chemical Industry, 45-70, also Proc. of 3rd Int. Conf. of the Centre for High Energy Forming, 1.3.1-1.3.43.

96. Wittman, R.H. (1973), The Influence of Collision Parameters on the Strength and Microstructure of an Explosion Welded Aluminium Alloy, Proc. 2nd Sym. on Use of Explosive Energy in Manufacturing Metallic Materials of New Properties and Possibilities of Application thereof in the Chemical Industry, 153-168.

97. Carpenter, S.H. and Wittman, R.H. (1974), Recent Developments in the Theory and Application of Explosive

Welding, Soc. of Manufacturing Engr., Tech. paper MF 74-819, 1-17.

98. Loyer, A., Hay, D.R. and Gagnon, G. (1975), Weldability Windows and the Selection of Explosive Welding Process Parameters, Proc. 5th Int. Conf. on High Energy Rate Fabrication, 4.3.1-4.3.14.

99. Loyer, A., Talerman, M., Hay, D.R. and Gagnon, G. (1976), Explosive Welding - The Weldability Window for Dissimilar Metals and Alloys, Proc. 3rd Int. Sym. on Use of Explosive Energy in Manufacturing Metallic Materials of New Properties and Possibilities of Application thereof in the Chemical Industry, 43-57.

100. Deribas, A.A., Simonov, V.A. and Zakcharenko, I.D. (1975), Investigation of Explosive Welding Parameters for Arbitrary Combinations of Metals and Alloys, Proc. 5th Int. Conf. on High Energy Rate Fabrication, 4.1.1-4.1.24.

101. Stivers, S.W. and Wittman, R.H. (1975), Computer Selection of the Optimum Explosive Loading and Weld Geometry, Proc. 5th Int. Conf. on High Rate Fabrication, 4.2.1-4.2.16.

102. Takizawa, Y. and Izuma, T. (1976), An Experiment on the Calculation of Explosive Bonding Conditions, Proc. 3rd Int. Sym. on the Use of Explosive Energy in Manufacturing Metallic Materials of New Properties and Possibilities of Application thereof in the Chemical Industry, 15-32.

103. Crossland, B. (1976), Review of the Present State-of-the-Art in Explosive Welding, Metals Technology, 3, 8-20.

104. Szecket, A. and Crossland, B. (1977), The Design and Development of a 63.5 mm Bore Gas Gun for Oblique Impact Experiments and Preliminary Results, Proc. 6th Int. Conf. High Energy Rate Fabrication, 7.2.1-7.2.15.

105. Meyer, M.D. (1973), Impact Welding using Magnetically Driven Flyer Plates, Proc. 4th Int. Conf. of the Centre for High Energy Forming, 5.3.1-5.3.23.

106. Szecket, A. (1979), An Experimental Study of the Explosive Welding Window, Ph.D. Thesis of the Queen's University of Belfast.

107. Dobrushin, L.D. (1979), A Problem of the Lower Limit of Explosion Welding, Avt. Svarha, No. 6, 64-65.

108. Vaidyanath, L.R., Nicholas, M.G. and Milner, D.R. (1959), Pressure Welding by Rolling, Brit. Weld. J. 6, 13-28.

109. McKenney, C.R. and Banker, J.G. (1971), Explosion-Bonded Metals for Marine Structural Applications, Marine Technology, 8, 285-292.

110. Schetky, L. McDonald (1976), Copper Hulls for Large Ships, Marine Week, Supplement on Marine Design, 3, 15 and 31.

111. Enright, T.J., Sharp, W.F. and Bergmann (1970), Explosive Bonding of Dissimilar Metals, Metal Progress, 98/1, 107-114.

112. Persson, P.I. (1976), Explosive Clad Electrode Material for the Electrochemical Industry and Off-shore Oil Platforms, Proc. 3rd Int. Sym. on the Use of Explosive Energy in Manufacturing Metallic Materials of New Properties by Explosive Welding, Cladding, Hardening and Metal-Powder-Pressing Processes, Mariánské Lázne, 193-204.

113. Takizawa, Yu. (1978), Private Communication.

114. Manganello, S.J. and Abbott, K.H. (1972), Ballistic Behaviour of Steel Targets, J. of Mat., 7, 231-239.

115. Crossland, B., Cave, J.A., Banerjee, S.K. and Wylie, H.K. (1975), Explosive Cladding of Large and Relatively Thick Flyer Plates, Proc. 5th Int. Conf. on High Energy Rate Fabrication, 4.10.1-4.10.23.

116. Persson, P.I. (1970), Experience from the Large Scale Commercial Production of Explosively Clad Compound Materials, Proc. 1st Int. Sym. Use of Explosive Energy in Manufacturing Metallic Materials of New Properties and Possibilities of Application thereof in Chemical Industry, 121-124.

117. Chudzic, B. (1963), Methods of Explosively Bonding Metals, British Patent 13 169 163.

118. Ruppin, D. (1968), The Explosion Welding of Metals - Investigation into the Unstable Processes Associated with the Movement of the Cover Plate, Colloquium on Welding by Thermochemical or Mechanical Energy, IIW meeting at Warsaw.

119. Wylie, H.K. and Crossland, B. (1973), Explosive Cladding with Thick Flyer Plates, Proc. Int. Conf. on the Use of

High-Energy Rate Methods for Forming, Welding and Compaction, Leeds, 15.1-15.21.

120. Bergmann, O.R., Buchwald, J. and Cowan, G.R. (1969), Process for Initiating Explosive and Charge Therefor, U.S. Patent 3 439 408.

121. Brennecke, H.M. (1965), Explosive Bonding, U.S. Patent 3 205 574.

122. Saia, J.J. (1968), Process for Explosion-Bonding Metals, U.S. Patent 3 360 848.

123. Crossland, B., Bahrani, A.S. and Shribman, V. (1969), Improvements in and Relating to the Bonding Together of Two Bodies, British Patent 1 249 299.

124. Sakhnovskaya, E.B., Sedykh, V.S. and Trykov, Yu.P. (1971), Properties of Explosion Welded Joints between Austenitic Steel and Aluminium Alloys, Svar Proiz, No. 7, 34-36.

125. Sakhnovskaya, E.B., Sedykh, V.S. and Trykov, Yu.P. (1972), Structural and Mechanical Inhomogeneity of Explosive-Welded Steel Aluminium Joints, Svar Proiz, No. 9, 7-9.

126. Teslenko, T.S., Sabolenko, T.M. and Berdichevskii, G.V. (1972), The Structure of Explosion Welded Joints, Avt. Svarka, No. 1, 1-4.

127. Holtzman, A.H. (1970), Hot Rolling Explosion-Bonded Stainless Steel/Carbon Steel Clads, U.S. Patent 3 489 618.

128. De Maris, J.L. (1967), Rolling Explosion-Bonded Titanium Clads, U.S. Patent 3 331 121.

129. Holtzman, A.H. (1966), Process of Treating Explosively Clad Metals, U.S. Patent 3 238 071.

130. Williams, C.P. (1970), Explosion Clad Titanium/Steel Plate, Chemical Engineering Progress, 66, 48-52.

131. Bouckaert, G.P., Hix, H.B. and Chelius, J. (1975), Explosive-Bonded Tantalum-Steel Vessels, Proc. 5th Int. Conf. on High Energy Rate Fabrication, 4.4.1-4.4.25

132. Ruppin, D. (1973), Process of Producing Inside-Plated Bottoms of Tanks by Explosive Welding. Proc. of the 2nd Int. Sym. on Use of Explosive Energy in Manufacturing Metallic Materials of New Properties, 141-151.

133. Phillipchuk, V. (1961), Welding, Forging and Cutting with Explosives, ASTME, 61, Paper No. 350.

134. Holtzman, A.H. and Rudershausen, C.G. (1962), Recent Advances in Metal Working with Explosives, Sheet Metal Industries, 39/421, 399-410.

135. Popoff, A.A. (1964), Explosive Welding Process, U.S. Patent 3 140 537.

136. Buckwald, J. (1971), Explosive Bonding of Tubes, U.S. Patent 3 562 897.

137. Barrington, J. (1968), Explosive Tube Bonding, U.S. Patent 3 364 561.

138. Dalrymple, D.G. and Johnson, W. (1967), Explosive Cladding of Thin Steel Tubes with Copper: and Exploratory Study, Int. J. Mach. Tool. Des. Res., 7, 257-267.

139. Doherty, A.E. and Knop, L.H. (1969), Practical Application of Explosive Welding, Proc. 2nd Int. Conf. of the Centre for High Energy Forming, 7.4.1-7.4.31.

140. Addison, H.J., Kowalick, J.F. and Cavell, W.W. (1969), Explosive Welding of Cylindrical Shapes, Eng. Conf. of ASTME paper, AD, 69-181.

141. Schetky, L. McDonald, (1970), The Explosive Fabrication of Duplex Tube, J. Inst. Metals, 98, 364-367.

142. Blazynski, T.Z. and Dara, A.R. (1970), The Use of Explosives for Implosive Welding of Duplex Cylinders, Proc. 11th Int. M.T.D.R. Conf., 941-956.

143. Blazynski, T.Z. and Dara, A.R. (1971), A Comparison between the Implosive and Explosive Systems for Welding of Duplex Cylinders, Proc. 3rd Int. Conf. of the Centre for High Energy Forming, 8.3.1-8.3.26.

144. Dara, A.R. and Blazynski, T.Z. (1973), Explosive Welding Systems for the Manufacture of Compound Cylinders, Proc. Int. Conf. on the Use of High Energy Rate Methods for Forming, Welding and Compaction, Leeds, 10.1-10.3.

145. Crossland, B., Wylie, H.K., Williams, P.E.G. and Bahrani, A.S. (1973), Explosive Welding of Cylindrical Surfaces, 2nd Int. Sym. Use of Explosive Energy in Manufacturing Metallic Materials of New Properties, 97-133.

146. Yoblin, J.A. and Mote, J.D. (1976), Explosion Welding-Some History; Current Processes and Applications. 3rd Int. Sym. Use of Explosive Energy in Manufacturing Metallic Materials of New Properties by Explosive Welding,

161-180.

147. Horeth, M. (1957), Method of Permanently Connecting Metal Pipes, British Patent 866 741.

148. Otto, H.E. and Carpenter, S.H. (1971), High Energy Rate Swaging of Aluminium Pipe Connectors, 3rd Int. Conf. of the Centre for High Energy Forming, 8.1.1-8.1.17.

149. Capper, H.M. and Chaplin, F.S. (1971), Development of a Propellant Actuated Pipe Joining System, 3rd Int. Conf. of the Centre for High Energy Forming, 8.2.1-8.2.23.

150. Willis, J. and Murdie, D.C. (1962), Explosive Joining of Aluminium-Alloy Pipes, Sheet Metal Industries, 39 811-813.

151. Verzeletti, G., Montagnani, M., Beghi, G. and Golinelli, P. (1968), Explosive Welding of Nuclear Reactor Materials, Select Conf. on Explosive Welding, pub., Welding Inst., 1-8.

152. Grollo, R.D. (1971), Explosive Joining of Dissimilar Metal Tubes, 3rd Int. Conf. of the Centre for High Energy Forming, 8.4.1-8.4.18.

153. Van der Horst, H. (1968), Welding by Pressure, U.K. Patent 1 220 171.

154. Chadwick, M.D. and Evans, N.H. (1973), Explosive Welding, Metal Construction, 5, 285-292.

155. Chadwick, M.D. and Evans, N.H. (1973), Explosive Welding Techniques for Joining Steel Pipes, Proc. Int. Conf. on the Use of High Energy Rate Methods for Forming, Welding and Compaction, Leeds, 13.1-13.7.

156. Dubin, A.A., Ermolaev, V.V., Lysenko, D.N., Orlov, B.D., Varlamov, G.I. and Karpukhin, V.F. (1971), Tube Welding with Pulsed Magnetic Field, Svar Proiz, 2, 16-18.

157. Howell, W.G., Espinoza, T.A. and Wittman, R.H. (1974), U.S. Patent 3 806 020.

158. Chadwick, M.D. (1976), Pipe to Pipe Welding by Explosives, Explosive Welding, pub. The Welding Institute, 31-39.

159. Wylie, H.K. and Crossland, B. (1973), The Explosive Welding of Tubes to Collars, 4th Int. Conf. of the Centre for High Energy Forming, 4.2.1-4.2.24.

160. Mottram, R.A. (1968), Explosive Tube to Tubeplate Expansion, Proc. of the HPTA Meeting on Use of Explosives

in forming, Welding and Compaction, Belfast, Paper 7.

161. Berman, I., Kunsagi, L., and Thakkar, B. (1967), Development and Use of Explosive Forming for Commercial Application, 1st Int. Conf. of the Centre for High Energy Forming, 3.1.1-3.1.45.

162. Berman, I. and Schroeder, J.W. (1973), Development, Use and Evaluation of Explosively Formed Components. 4th Int. Conf. of the Centre for High Energy Forming, 1.4.1-1.4.15.

163. Schroeder, J.W. and Berman, I. (1977), Detnaforming of Feedwater Heater Tubes at Power Plant Installations, 6th Int. Conf. on High Energy Rate Fabrication, 2.6.1-2.6.25.

164. Brown, D.J. and Dunmore, O.J. (1973), Explosive Expansion of Tubes into Tubeplates, Proc. Int. Conf. on the Use of High-Energy Rate Methods for Forming, Welding and Compaction, Leeds, 19.1-19.8.

165. Crossland, B., Bahrani, A.S., Williams, J.D. and Shribman, V. (1967), Explosive Welding of Tubes to Tubeplates, Welding and Metal Fabrication, 35, 88-94.

166. Yorkshire Imperial Metals Ltd., (1969), Improvements in Securing Tubes into Tubeplates, British Patent 1 149 387.

167. Associated Electrical Industries Ltd., (1970), Securing a Tube in an Opening in a Metal Plate, British Patent 1 179 107.

168. Cairns, J.H. and Hardwick, R. (1968), Explosive Welding of Tubes into Tubeplates Using Angular Geometry, Select Conf. on Explosive Welding, pub. Welding Institute, 67-72.

169. Cairns, J.H. and Hardwick, R. (1971), Explosive Welding of Titanium Tubes to Tubeplates by the YIMpact Process, 3rd Int. Conf. of the Centre for High Energy Forming, 2.3.1-2.3.19.

170. Hardwick, R. (1973), The Application of Explosive Welding to Industrial Heat Exchangers, Proc. Int. Conf. on the Use of High-Energy Rate Methods for Forming, Welding and Compaction, Leeds, 2.1-2.10.

171. Hardwick, R. (1973), The YIMpact Explosive Tube to Tubeplate Welding System, 2nd Int. Conf. on Uses of Explosive Energy in Manufacturing Metallic Materials of New Properties, Mariánske Lázne, 185-198.

172. Hardwick, R. (1975), Factors Influencing the Reduction of Ligament Thickness when Welding Tube to Tubeplate Joints, Explosive Welding, pub. Welding Institute, 12-18.

173. Shribman, V., Williams, J.D. and Crossland, B. (1968), Some Problems of Explosively Welded Tubes to Tubeplates, Select Conf. on Explosive Welding, pub. Welding Institute, 47-54.

174. Crossland, B. and Williams, P.E.G. (1973), An Experimental Investigation of the Permissable Ligament Thickness in Tube to Tubeplate Welding, Proc. Int. Conf.on Use of High-Energy Rate Methods for Forming, Welding and Compaction, Leeds, 9.1-9.13.

175. Otto, H.E., Carpenter, S. and Pfluger, A. (1977), Explosive Welding and Propellant Swaging of Tubes to Tubesheets, 6th Int. Conf. of High Energy Rate Fabrication, 2.8.1-2.8.23.

176. Prümmer, R. and Jähn, F. (1977), Improved Method of Explosion Welding Tubes to Tubesheet, 6th Int. Conf. of High Energy Rate Fabrication, 2.7.1-2.7.15.

177. Johnson, W.R. (1971), Explosive Welding Plugs into Heat Exchanger Plates, Welding Journal, 50, 22-32.

178. Bahrani, A.S., Halliburton, R.F. and Crossland, B. (1972), The Explosive Plugging of Heat Exchangers, Proc. Int. Conf. on Welding Research Related to Power Plant, 617-633.

179. Crossland, B., Bahrani, A.S. and Townsley, W.J. (1977), Explosive Plugging of Nuclear Heat Exchangers, Proc. 3rd Int. Conf. on Pressure Vessel Technology, 971-983.

180. Bahrani, A.S., Halliburton, R.F. and Crossland, B. (1973), Parallel Technique of Tube to Tubeplate Welding Applied to Plugging of Heat Exchangers, The Int. Journal of Pressure Vessels and Piping, 1, 17-35.

181. Dalrymple, D.G. (1972), Current Canadian Use of Explosive Welding for Repair and Manufacture of Nuclear Steam Generators, AECL Research and Development in Engineering, Winter issue, 17-21.

182. Hardwick, R. (1975), Explosive Plugging and Factors which have Influenced Commercial Viability, Explosive Welding, pub. Welding Institute, 28-30.

183. Crossland, B., Braithwaite, A.B.M. and Chapman, H.E. (197

Explosive Plugging Development for Nuclear Heat Exchangers, Proc. of Conf. of BNES; Welding and Fabrication in the Nuclear Industry, London, 121-127.

184. Jackson, P.W. (1975), Explosive Plugging of Heat Exchangers on Site, Explosive Welding, pub. Welding Institute, 24-26.

185. Jackson, P.W. and Brown, D.J. (1980), Explosive Welding, IRD Newsletter, No. 38.

186. Addison, H.J. (1964), Explosive Welding, ASME paper 64-MD-47.

187. Kogya, A.K. and Kaisha, K. (1966), Line Welding of Metals by Explosives, U.K. Patent 1 046 655.

188. Polhemus, F.C. (1967), Explosive Welding Development at Pratt and Whitney Aircraft, 1st Int. Conf. of the Centre for High Energy Forming, 1.3.1-1.3.17.

189. Bement, L.J. (1972), Small-Scale Explosive Seam Welding, Presented at a Symposium on Welding, Bonding, and Fastening at Williamsburg, also issued by NASA, Langley Research Centre.

190. Fukuyama, I. (1970), Explosive Working of Metals in Japan, 1st Int. Conf. on Uses of Explosive Energy in Manufacturing Materials of New Properties, Mariánské Lázne, 353-357.

191. Willis, J. (1975), Application of Explosive Welding, Explosive Welding, pub. Welding Institute, 40-44.

192. Unterlerchner, W. and Prümmer, R. (1977), Production and Properties of Tubular Transition Joints, 6th Int. Conf. on High Energy Rate Fabrication, 2.13.1-2.13.14.

193. Wright, E.S. and Bayce, A.E. (1964), Current Methods and Results in Explosive Welding, NATO Advanced Study Institute Meeting on High Energy Rate Working of Metals, Norway, 448-467.

194. Izuma, T. and Baba, N. (1977), Development of Transition Joint for Cryogenic Temperature, 6th Int. Conf. on High Energy Rate Fabrication, 2.14.1-2.14.19.

195. Holtzman, A.H. and Cowan, G.R. (1965), Bonding of Metals with Explosives, Welding Research Council Bulletin, No. 104.

196. Wylie, H.K. and Crossland, B. (1977), A Feasibility Study of the Explosive Fabrication of Triplex Tubing with Axial

Channels at the Interface between the Component Tubes. 6th Int. Conf. of High Energy Rate Fabrication, 2.4.1-2.4.10.

197. Velten, R. (1973), Practical Applications of Explosive Welding, 4th Int. Conf. of the Centre for High Energy Forming, 8.4.1-8.4.28.

198. Persson, I. (1973), Explosive Welding Indoors in Serial Production, 2nd Int. Sym. on Use of Explosive Energy in Manufacturing Metallic Materials of New Properties, 261-270.

199. Kameishi, M., Baba, N., Niwatsukino, T. and Fukumoto, C. (1976), Present State of Explosive Cladding in Japan, 3rd Int. Sym. on Use of Explosive Energy in Manufacturing Metallic Materials of New Properties, 205-223.

200. Erdmann-Jesnitzer, F., Louis, H. and Wiedemeier, J. (1977), Spot Welding of Thin Metal Foils with the Fluid Impact Welding Process, Machinery Market, 85, 1907-1909, also (1978), Fluid Impact Welding-Explosion Cladding, IIW Document IV 248-78.

201. Jarvis, C.V. and Slate, P.M.B. (1968), Explosive Fabrication of Composite Materials, Nature, 220, 782-783.

202. Slate, P.M.B. and Jarvis, C.V. (1972), Strengthening of Metals by Explosive Incorporation of Strong Wires, J.Inst. of Metals, 100, 217-224.

203. Fleck, J., Laber, D. and Leonard, R. (1969), Explosive Welding of Composite Materials, J. Composite Mats., 3, 699-701.

204. Reece, O.Y. (1970), Explosive Bonding Packs Strength into Metal Composites, Iron Age, 205 pt., 60-61.

205. Reece, O.Y. (1971), Molybdenum Wire Reinforced Columbium Composites, Proc. 3rd Int. Conf. of the Centre for High Energy Forming, 2.1.1-2.1.11.

206. Wylie, H.K., Williams, J.D. and Crossland, B. (1971), Fabrication of Metal/Metal Wire Composite Materials by the Use of Explosives, Welding and Metal Fabrication, 39, 214-219.

207. Kelly, A. (1965), The Fibre Reinforcement of Metals, pub. HMSO.

208. Kelly, A. and Tyson, W.R. (1965), Tensile Properties of

Fibre Reinforced Metals: Copper/Tungsten and Copper Molybdenum, J. of the Mech. and Phys. of Solids, 13, 329-350.

209. Wylie, H.K. (1971), Explosive Welding; Parameters and Applications, Ph.D. Thesis, Queen's University of Belfast.

210. Bhalla, A.K. and Williams, J.D. (1977), Production of Stainless Steel Wire-Reinforced Aluminium Composite Sheet by Explosive Compaction, J. Mat. Sci., 12, 522-530.

211. El-Sobky, H. and Blazynski, T.Z. (1973), A Preliminary Investigation of the Implosive Welding of Filament Metal-Foil Composite Cylinders, Proc. Int. Conf. on the Use of High Energy Rate Methods for Forming Welding and Compaction, Leeds, 12.1-12.8.

212. Blazynski, T.Z. and El-Sobky, H. (1975), Explosive Welding of Multi-layer, Wire-Mesh Reinforced, Metal-Foil Cylinders, Proc. 5th Int. Conf. on High Energy Rate Fabrication, 4.6.1-4.6.24.

213. Blazynski, T.Z. and El-Sobky, H. (1976), Parameters of Implosive Welding of Multilayered, Mesh Reinforced Foil Cylinders, Proc. 3rd Int. Sym. on Use of Explosive Energy in Manufacturing Metallic Materials of New Properties, 351-365.

214. Haegland, B. (1967), Applications of High Energy Rate Forming in Norway, Proc. 1st Int. Conf. of the Centre for High Energy Forming, 2.6.1-2.6.26.

215. Blazynski, T.Z. and Bedroud, Y. (1977), Properties of Implosively Welded Co-Axial Multi-Strand Composites, Proc. 6th Int. Conf. on High Energy Rate Fabrication, 2.12.1-2.12.11.

216. Standard Specification for Corrosion-Resisting Chromium Steel Clad Plate, Sheet and Strip, American National Standard, ANSI/ASTM A 263-77.

217. Standard Specification for Stainless Chromium-Nickel Steel Clad Plate, Sheet and Strip, American National Standard, ANSI/ASTM A 264-77.

218. Standard Specification for Nickel and Nickel-Base Alloy Clad Steel Plate, American National Standard, ANSI/ASTM A 265-77.

219. Standard Specification for Copper and Copper Alloy Clad Steel Plate, American National Standard, ANSI/ASTM B 432-76.

220. Bahrani, A.S. and Crossland, B. (1966), Further Experiments on Explosive Welding and Cladding with Particular Reference to the Strength of the Bond, Proc. Inst. Mech. Engr., 180, 31-46.

221. Frey, A.P. (1968), Some Methods of Quality Testing for Explosive Welded Metal Sheets, Select Conf. on Explosive Welding, pub. Welding Institute, 63-65.

222. Anderson, D.K.C. (1977), The Evaluation of Explosively Clad Plates by Destructive and Non-Destructive Testing, 6th Int. Conf. on High Energy Rate Fabrication, 1.12.1-1.12.8.

223. De Maris, J.L. and Pocalyko, A. (1966), The Mechanical Properties of Detaclad Explosion-Bonded Clad Metal Composites, 1966 ASTME Engineering Conference Paper AD 66-113.

224. Gelman, G.S., Zairsev, G.Z., Ponomarev, V.Ya. and Tsemakhovich, B.D. (1966), Fatigue Strength of Two Layer Steel, Clad by Explosion Bonding, Svar. Proiz., No. 10, 4-6.

225. Banerjee, S.K. and Crossland, B. (1971), Mechanical Properties of Explosively-Cladded Plates, Metal Construction and British Welding Journal, 3, 268-273.

226. Avery, H.S. 1959, The Mechanism of Thermal Fatigue: Metal Progress, 76-Aug., 67-70

227. Baron, H.G. (1964), Thermal Shock and Thermal Fatigue, Thermal Stress, pub. Pitman, Chapter 10, 182-206.

228. Kimura, K. and Takahashi, S. (1977), Ultrasonic Examination of Clad Steel Bond with Dissimilar Acoustic Impedance, Proc. 3rd Int. Conf. on Pressure Vessel Technology, 449-456.

229. Robinson, P., Abrahams, C.J. and Gibson, R. (1968), Ultrasonic Examination of Explosively Welded Tube to Tubeplate Joints, Select Conf. on Explosive Welding, pub. Welding Institute, 39-45.

230. Taksukawa, I. (1970), Residual Stress Measurements on Explosive Clads, J. Jap. Weld. Soc., 39, No. 9, 61-67.

231. Taksukawa, I. and Oda, I. (1971), Residual Stress Measurements on Explosive Stainless Steel Clad Steel, Trans. Jap. Weld. Soc., 2, No. 2, 26-34.

232. Pruemmer, R.A. (1971), Metallurgical, Microradiographic and X-Ray Residual Stress Investigation of Explosively Welded Aluminium-Steel, 3rd Int. Conf. of the Centre for High Energy Forming, 3.5.1-3.5.16.

233. Pathiraj, B. and Vasudevan, R. (1977), X-Ray Residual Stress Measurement on Explosively Welded Aluminium-Mild Steel Weld, 6th Int. Conf. on High Energy Rate Fabrication, 1.14.1-1.14.12.

234. Lucas, W. and Williams, J.D. (1968), Some Metallurgical Aspects of Explosive Welding, Proc. of the High Pressure Technology Association on the Use of Explosives in Forming, Welding and Compaction, Paper 9.

235. Williams, J.D., Dhir, P. and Crossland, B. (1971), The Nature of Vortices in Explosive Welds, Proc. 3rd Int. Conf. of the Centre for High Energy Forming, 3.1.1-3.1.22.

236. Dhir, P., Williams, J.D. and Crossland, B. (1973), Non-Equilibrium Alloy Formation in Explosive Welds, Proc. 4th Int. Conf. of the Centre for High Energy Forming, 2.1.1-2.1.25.

237. Loosemore, G.R. and Peapell, P.N. (1979), How Explosive Welding Affects Carbon Steel Plate, Welding and Metal Fabrication, 47, 229-236.

238. Smith, C.S. (1958), Metallographic Studies of Metals after Explosive Shock, Trans. Met. Soc. AIME, 212, 574-589.

239. Smith, C.S. and Fowler, C.M. (1960), Further Metallurgical Studies on Metals after Explosive Shock, Proc. Met. Soc. Conf. Response of Metals to High Velocity Deformation, 309-341.

240. Zukas, E.G. and Fowler, C.M. (1960), The Behaviour of Iron and Steel under Impulsive Loading, Proc. Met. Soc. Conf. Response of Metals to High Velocity Deformation, 343-369.

241. Tardif, H.P., Claisse, F. and Challet, P. (1960), Some Observations on Explosively Loaded Iron and Mild Steel, Proc. Met. Soc. Conf. Response of Metals to High Velocity

Deformation, 389-407.

242. Dieter, G.E. (1960), Metallurgical Effects of High Intensity Shock Waves in Metals, Proc. Met. Soc. Conf. Response of Metals to High Velocity Deformation, 409-445.

243. Holtzman, A.H. and Cowan, G.R. (1960). The Strengthening of Austenitic Manganese Steel by Plane Shock Waves, Proc. Met. Soc. Conf. Response of Metals to High Velocity Deformation, 447-482.

244. Appleton, A.S., Dieter, G.E. and Bever, M.B. (1961), Effects of Explosive Shock Waves on Gold-Silver Alloy, Trans. Met. Soc. AIME, 221, 90-94.

245. Curran, D.R., Kratz, S., Kelly, J.J. and Nicholson, M.E. (1959), Pressure Required for Transformation Twinning in Explosively Loaded Low-Carbon Steel, Trans. Met. Soc. AIME, 215, 151-152.

246. Cohen, J.B., Nelson, A. and De Angelis, R.J. (1966), Observations of some Shock Loaded Copper, Trans. Met. Soc. AIME, 236, 133-135.

247. Bowden, H.G. and Kelly, P.M. (1967), The Structure and Mechanical Properties of Shock-Loaded Steels, Metal Sci. J., 1, 75-77.

248. Brillhart, D.C., De Angelis, R.J., Preben, A.G., Cohen, J.B. and Gordon, P. (1967), Quantitative Study of the Substructure and Properties of Shock Loaded Copper, Trans. Met. Soc. AIME, 239, 836-843.

249. Otto, H.E. and Mikesell, R. (1967), The Shock Hardening of Structural Metals, Proc. 1st Int. Conf. of the Centre for High Energy Forming, 2, 7.6.1-7.6.46.

250. Johari, O. and Thomas, G. (1964), Factors Affecting Dislocation Sub-Structure in Deformed Copper, Acta Met., 12. 679-686.

251. Weertman, J., (1960), High Velocity Dislocations, Proc. Met. Soc. Conf. Response of Metals to High Velocity Deformation, 205-247.

252. O'Brien, J.L. and Davis, R.S. (1960), On the Fracture of Solids under Impulsive Loading Conditions, Proc. Met. Soc. Conf. Response of Metals to High Velocity Deformation, 371-388.

253. Hayes, G.A. and Pearson, J. (1962), Metallurgical

Properties of Some Explosively Welded Metals, NAVWERS REPORT 7925.

254. Davenport, D.E. (1962), Explosive Welding, 1961-62 Advanced High Energy Rate Forming, Book II, SP 62-77.

255. Boes, P.J.M. (1962), Some Aspects of Explosive Welding, Tech. Centre for Metalworking T.N.O., Rep. 103.

256. Van Wely, F.E. and Verbraak, C.A. (1964), Material Aspects of High Energy Rate Deformation at Ambient and Elevated Temperatures, NATO Advanced Study Institute Conference on High Energy Rate Working of Metals, 172-190.

257. Carlson, R.J. (1965), Explosive Welding, Amer. Soc. Metals Tech. Rep. W6-3-65.

258. Deribas, A.A., Matveenkov, F., Sobolenko, T. and Teslenko, T. (1967), Investigations of the Changes in the Zones of Collision by Explosive Joining of Metals, Symposium High Dynamic Pressure, Paris.

259. Buck, G. and Hornbogen, E. (1964), Metallographic Investigation of Shock Welded Metals by Electron Microscopy, NATO Advanced Study Institute Conference on High Energy Rate Working of Metals, 517-534.

260. Buck, G. and Hornbogen, E. (1966), Metallographic Untersuchungen an ExplosivschweiBnähten, Metall. Tech., 20, 9-21.

261. Crossland, B. and Bahrani, A.S. (1967), A Review of Explosive Welding Research Carried out in the Queen's University of Belfast, Proc. 1st Int. Conf. of the Centre for High Energy Forming, 1, 1.1.1-1.1.50.

262. Pietteur, J. (1967), Industrial Applications of Steel Explosive Forming in Belgium, Methods and Development, Proc. 1st Int. Conf. of the Centre for High Energy Forming, 1, 2.1.1-2.1.68.

263. Berdichevskii, G.I. and Sobolenko, T.M. (1968), Research into the Zone of the Joint during Explosion Welding of Metals, Act. Svarka., No. 9, 13-17.

264. Trueb, L.F. (1968), An Electron Microscope Investigation of Explosion-Bonded Metals, Trans. Met. Soc. AIME, 242, 1057-1065.

265. Williams, J.D. (1969), Some Metallurgical Aspects of Explosive Welding, Ph.D. Thesis of the Queen's University

of Belfast.

266. Dhir, P. (1973), Metallurgy of Vortices in Explosion Welds, Ph.D. Thesis of the Queen's University of Belfast.

INDEX